Environmental Biochemistry

Environmental Biochemistry

Victor Perry

STATES
ACADEMIC PRESS
www.statesacademicpress.com

States Academic Press,
109 South 5th Street,
Brooklyn, NY 11249, USA

Visit us on the World Wide Web at:
www.statesacademicpress.com

ISBN: 978-1-63989-180-1 (Hardback)

Cataloging-in-Publication Data

Environmental biochemistry / Victor Perry.
p. cm.
Includes bibliographical references and index.
ISBN 978-1-63989-180-1
1. Biochemistry--Environmental aspects. 2. Environmental chemistry.
3. Biochemistry. 4. Microbial ecology. I. Perry, Victor.

QD415 .E58 2022
572--dc23

Table of Contents

Preface

The purpose of this book is to help students understand the fundamental concepts of this discipline. It is designed to motivate students to learn and prosper. I am grateful for the support of my colleagues. I would also like to acknowledge the encouragement of my family.

The study of chemical processes related to and within living organisms is known as biochemistry. Environmental biochemistry aims to examine the biochemical processes and its effects of various organic and inorganic compounds present in the environment. The application of principles of biochemistry for the preservation of the environment is the underlying goal of this field. It is concerned with the study of microbial metabolism of pollutants with a focus on understanding metabolites and critical metabolic reactions. It is also concerned with analyzing enzymes involved in metabolism of pollutants based on structural, genetic and kinetic criteria. Protection from radiation, managing air resources and water quality and preserving industrial hygiene are some of the areas of study under this discipline. The book aims to shed light on some of the unexplored aspects of this field. It presents the complex subject of environmental biochemistry in the most comprehensible and easy to understand language. This book, with its detailed analyzes and data, will prove immensely beneficial to professionals and students involved in this area at various levels.

A foreword for all the chapters is provided below:

Chapter - Introduction

The scientific study of the chemical and biochemical phenomena that occur in natural places is called environmental chemistry. Environmental biochemistry is concerned with the study of microbial metabolism of pollutants, metabolites and critical metabolic reactions. This chapter has been carefully written to provide an easy understanding of environmental biochemistry.

Chapter - Atmospheric and Water Chemistry

The branch of atmospheric science that is concerned with the chemical processes occurring within the Earth's atmosphere is called atmospheric chemistry. Water chemistry studies chemical components and properties of water. This is an introductory chapter which will briefly introduce about atmospheric and water chemistry.

Chapter - Soil Biochemistry

Soil biochemistry is a branch of soil science that is associated with the formation and decomposition of soil organic matter, biochemical reactions of carbon, nitrogen, phosphorus, sulfur, metals and xenobiotics in soils, etc. This chapter closely examines the aspects of soil biochemistry to provide an extensive understanding of soil biochemistry.

Chapter - Ecological Biochemistry

The chemical interaction of organisms with their abiotic environment is called ecological biochemistry. It further includes chemical ecology, pesticide biochemistry, pesticide chemistry, etc. This chapter discusses the subject of ecological biochemistry in detail.

Chapter - Green Chemistry

The branch of chemistry and chemical engineering that focuses on designing of processes and products for mitigating the use of hazardous substances is termed as green chemistry. A few of its aspects are green chemistry metrics and principles, etc. The topics elaborated in this chapter will help in gaining a better perspective of green chemistry.

Victor Perry

1

Introduction

The scientific study of the chemical and biochemical phenomena that occur in natural places is called environmental chemistry. Environmental biochemistry is concerned with the study of microbial metabolism of pollutants, metabolites and critical metabolic reactions. This chapter has been carefully written to provide an easy understanding of environmental biochemistry.

Environment

The environment refers to our immediate surroundings in which all living and non-living components co-exist. It can be used to refer to the social, natural or built environment. The environment has a huge impact on an individual's behaviour and intellect. Any change that occurs in a living entity either directly or indirectly is related to a change in his immediate surrounding.

Types of Environment

There are two different types of environment:

- Geographical Environment.
- Man-made Environment.

Geographical Environment

It consists of all components provided by nature and hence can be called as the natural environment. It is also referred to as the physical environment as it pertains to the physical requirements of life. These physical or geographic conditions are not dependent on the existence of humans. Sometimes, humans have no control over the physical conditions of the environment.

It includes natural resources, the earth's surface, mountains, plains, land, water, deserts, storms, cyclones, volcanoes, oceans, climatic factors, and so on. It is also used to refer to biological situations such as complexities associated with plants and animals. The sustainability of the natural resources is known to contribute towards the economy of a country.

Man-made Environment

This environment is used to refer to the one created by man in order to regulate and monitor certain environmental conditions. Some address it as a social-cultural environment. It can further be divided into two types of environments.

- Inner Environment.
- Outer Environment.

The Inner Environment

It is a social environment and it exists as long as a particular society exists. It pertains to the regulations, traditions, organizations and institutions. It involves customs and folkways which is existent in every human group. It is addressed with names such as non-material culture, social heritage etc. This heritage is essential for the social life of humans to flourish, it is known to have an influence on an individual's life. The altered form of the economic and physical environment – artificial environment, are seen as two different aspects of the man-made environment.

The Outer Environment

Through advancement in the field of science and technology, humans have attempted to alter conditions of their physical environment. This outer environment is as a result of these modifications which includes modern infrastructure in cities, our homes and their associated amenities, our modes of communication and transport, our resorts to conveniences and luxury, different kinds of industry manufacturing luxurious commodities, electrical appliances and so on which ultimately aims at civilization and urbanization. The inner and the outer environments are correlated and hence inseparable.

Environmental Chemistry

Environmental chemistry is the discipline that concerns itself with how chemicals are formed, how they are introduced into the environment, how they change after being introduced, the extent to which they enter and where they end up in organisms and other receptors, and the effects they have (usually the damage they do) once they get there.

Physical Chemistry in the Environment

Inorganic chemical reactions and organic chemical reactions are certainly an important aspect of environmental chemistry but there is another aspect of chemistry that must not be ignored, and this is physical chemistry.

When a chemical is released into the environment, it becomes distributed among the four major environmental compartments: (1) Air, (2) water, (3) soil, and (4) flora and fauna, that is, living organisms. Each of the first three categories can be further subdivided in floral (plant) environments and faunal (animal, including human) environments. The fraction of the chemical that will move into each compartment is governed by the physicochemical properties of that chemical.

In addition, the distribution of chemicals in the environment is governed by physical processes such as (1) sedimentation, (2) adsorption, and (3) volatilization and the chemical can then be degraded by chemical and biological processes. Chemical processes generally occur in water or the atmosphere and follow one of four reactions: oxidation, reduction, hydrolysis, and photolysis. Biological mechanisms in soil and living organisms utilize oxidation, reduction, hydrolysis, and conjugation to degrade chemicals. The process of degradation will largely be governed by the compartment (water, soil, atmosphere, biota) in which the chemical is distributed, and this distribution is governed by the physical processes already mentioned (i.e., sedimentation, adsorption, and volatilization).

The impact of the changes in the chemical state of organic chemicals on the environment is only partially elucidated, but will be significant in many cases. Atmospheric abundance of radioactive gases could lead to substantial drift in the climate of the earth, including changes in temperature and precipitation, and in the frequency of occurrence of extreme events (such as hurricanes). Ecosystem damage and problems related to human health also result from regional and global pollution, such as acid rain (also called *acidic precipitation*), which can vary between mildly acidic and strongly acidic on the pH scale and can suppress life and, together with enhanced ozone levels, can lead to forest damage. In fact, when assessing the impact of chemicals on the environment, the most critical characteristics are: (1) The type of chemical discharged, which depends on the type of industries and processes used, and (2) the amount and concentration of the chemical. Solid wastes (containing chemicals) and gaseous emissions generated from industrial sources also contribute to the amount and concentration of chemicals in the environment.

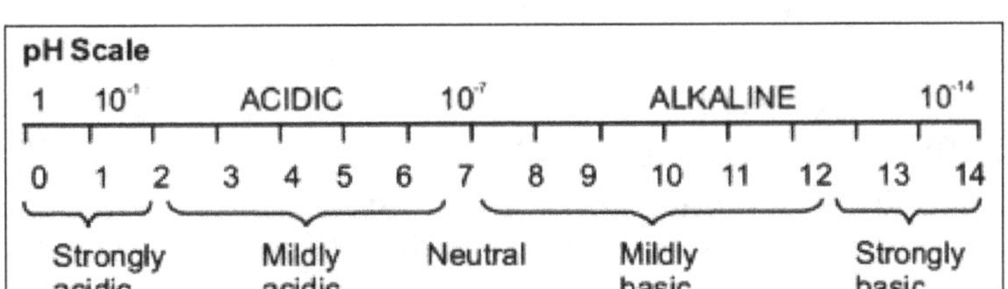

The pH scale.

Bioavailability

Bioavailability integrates environmental chemistry and ecotoxicologyconcepts. The bioavailable contaminant fraction represents the relevant exposure dose for organisms. Many complex processes describe bioavailable exposure dose, including the mass transfer and uptake of contaminants into organisms, which are determined by substance properties, compartment properties, the biology of organisms, and climatic influences. Expansion of bioavailabilityconsiderations into ecological risk-assessment decision tools will eventually improve decision making. Development of an ecological risk-assessment framework that incorporates bioavailability considerations can reduce uncertainty and provide better predictive power. However, the lack of validated analytical tools and models limits our current understanding of bioavailability. As bioavailable analytical tools and models continue to be developed and mechanistic-based understanding of bioavailable processes increases, confidence in the use of bioavailability will continue to increase.

Acid-base Chemistry

Environmental chemistry has a much different perspective on acid-base chemistry than the one you encountered in general chemistry. The most important difference is the *reference point* used to define acidity. General chemistry defines *acidity* and *basicity* relative to $a_{H^+} = 10^{-7}$ or pH = 7, the pH of pure water. Environmental chemists define *acidity* and *alkalinity* relative to $a_{H^+} = 10^{-5.5}$ or pH = 5.5, the pH of pure water in equilibrium with an atmosphere containing carbon dioxide gas $CO_2(g)$ at a partial pressure of $pCO_2 = 10^{-3.41}$ atm. The environmental reference point acknowledges the ubiquity and importance of aqueous $CO_2(aq)$ in atmospheric, surface and soil pore water.

The environmental perspective on acid-base chemistry is also strongly influenced by two seemingly conflicting features of environmental acidity and alkalinity: The capacity of natural waters to resist changes in pH yet remain susceptible to shifts resulting from human activity ranging from land use and waste management to atmospheric emissions. Compounds dissolved in water and chemical reactions on the surfaces of mineral colloids buffer the pH of natural water, yet this buffer capacity is finite and can be overwhelmed by acidity released by human activity. The magnitude and rate at which the pH of natural water decreases in response to the release of anthropogenic acidity determines the viability of natural ecosystems and alters human exposure to inorganic toxicants.

Basicity and Alkalinity

Environmental chemistry and the earth sciences have a much different perspective on acid-base chemistry than the one you encountered in general chemistry. The most important difference is the reference point used to define acidity and basicity. General chemistry defines acidity and basicity relative to $a_{H^+} = 10^{-7}$ mol dm^{-3} or pH 7; the pH of pure water.

Environmental chemists rely on the *alkalinity* concept to distinguish between solutions with an excess of strong bases (*alkaline* solutions) and solution with an excess of strong acids (*mineral acid* solutions). Unlike pure water where the reference point is pH 7, the reference point in natural waters is not a fixed pH value.

For instance, the mean atmospheric CO_2 (g) partial pressure for June 2015 at the Mauna Loa Observatory was $p_{CO2} = 4.028 \times 10^{-4}$ atm. A two-phase system consisting of pure water in equilibrium with a gas phase containing CO_2 (g) at the mean partial pressure recorded for June 2015 at the Mauna Loa Observatory would be pH 5.61. Pure chemists consider water saturated by CO_2 (g) acidic. Environmental chemists consider pure water saturated by CO_2 (g) as water lacking excess strong acids and excess strong bases, choosing to ignore the ever present dissolved CO_2 (g) when tallying mineral acidity and alkalinity.

Aqueous Carbon Dioxide

The National Oceanic and Atmospheric Administration (NOAA) began recording atmospheric monthly mean carbon dioxide contents at the Manua Loa Observatory in March 1958 and maintains an RSS feed of monthly mean atmospheric concentrations. Using the June 2015 mean monthly value of 402.80 ppm(v) CO_2 (g) we can simulated the pH of a two-phase gas-solution system[2] using *ChemEQL*.

The first step is apply Henry's Law to compute $c_{CO2}(aq)$ for $p_{CO2} = 4.0280 \times 10^{-4}$ atm.

$$H^{cp}_{CO_2} = \frac{a_{CO_2}(aq)}{f_{CO_2}} \approx \frac{\sim c_{CO_2(aq)}}{p_{CO_2}}$$

$$c_{Co_2(aq)} = \left(3.40 \times 10^{-2}\right) \cdot \left(4.0280 \times 10^{-4}\right)$$

$$c_{Co_2(aq)} = 1.3695 \times 10^{-5}\ \mathrm{mol\,dm^{-3}}$$

Applying the fixed-fugacity method requires component-specie CO_2(aq), mode(CO_2(aq)) = free and an initial CO_2(aq) concentration of: 1.37E−05.

The equilibrium pH of pure water in equilibrium with the above-ground atmosphere is 5.6.

Table: ChemEQL species or output table (bottom) for 1.3695×10^{-5} M carbon dioxide solution.

Species	Log(K)	Concentration (mol dm^{-3})	Activity (Unitless)
CO_2(aq)	0.00	1.37E−05	1.37E−05
H_2CO_3(aq)	−2.92	1.65E−08	1.65E−08
HCO_3^-	−6.37	2.38E−06	2.35E−06
CO_3^-	−16.70	4.64E−11	4.43E−11
OH^-	−14.00	4.07E−09	4.03E−09
H^+	0.00	2.48E−06	2.48E−06

The environmental interest in acid-base chemistry is focused on the capacity of natural waters and soils to resist pH changes resulting from human activity. Compounds dissolved in water and chemical reactions on the surfaces of mineral colloids buffer soil and surface water pH, yet this buffer capacity can be overwhelmed by acidity released by atmospheric deposition and soil fertility management. The magnitude and rate at which soil and surface water pH change in response to human activity influences ecosystem function and human exposure to inorganic toxicants.

Importance of Environmental Chemistry

Environmental chemistry deals with the effect of pollutants on the environment and the ways and means to reduce the contamination. This branch of chemistry is the study of pollutants behavior from the environmental point of view. The environmental chemistry is further classified into main areas; measurement of pollutants, and study of the behavior of the contaminants.

Environmental chemistry is vital as chemicals introduced into the environment proves to be harmful not only to the environment but also to human health and economy.

Importance of Environmental Chemistry for Human Health

Environmental chemistry helps to develop methods and procedures to reduce the contaminants or the chemicals in the air, which improves the quality of air. Cleaner air with fewer chemicals leads to less damage to the lungs. Several methods and preventive procedures have been developed with the help of environmental chemistry to reduce the release of hazardous substances into the water bodies, which leads to clean and recreational drinking water. The study also helped in reducing the risk for workers working in the chemical industries by inventing procedures which minimize the use of chemicals in manufacturing products.

Safer products with fewer chemicals are available for consumption purposes resulting in reduced chemical waste which results in less environmental pollution. This branch of chemistry also helps in introducing new ways of farming using fewer chemicals and more organic compounds to make safer food available for people across the world. It also helps to eliminate toxic chemicals from entering the food and food chain with the help of pesticides that are degradable and are toxic only to specific pests.

Importance of Environmental Chemistry for Environmental Safety

Many chemicals are introduced into the environment from various sources which include pesticides, harmful emissions from factories and vehicles, in the form of chemical waste from industries and many others.

Environmental chemistry researches into the core cause and develops methods, techniques and tools that reduce the chemical disposal into the environment. The chemistry has also contributed to developing green chemicals that either degrade compounds to innocuous products or to recover them for further use. Reduced chemicals in the environment help plants and animals to suffer less from harmful chemicals. It also helps to reduce global warming by lowering the rate of ozone depletion and pollutant deposits in thickly populated places. It also contributes a lot to less use of hazardous landfills that are persistent.

Importance of Environmental Chemistry for Economy and Business

Environmental chemistry is not only important to humans and the environment but is also crucial for business and economy. It helps in finding methods and techniques to speed up the chemical reactions using small amounts of reactants to deliver the same results and the same price of the product.

It also helps in reducing the number of synthetic steps which enables increased production, increase in the plant capacity while saving energy and water consumption. Using fewer chemicals for manufacturing the products results in reduced waste which results in reduced cost for disposing of chemical waste and toxic waste treatments. Environmental chemistry facilitates the replacement of new and purchased raw material with derivative waste material from other processes.

The environmental chemistry helps in developing methods and procedures which facilitates better performance by using less raw materials to achieve the desired results. It also helps in preventing the depletion of natural resources by reducing their use in the manufacturing industries. The greener products that are used by using environmental chemistry can generate more sales as they carry a safer product label.

There is a unique need for greener products, reducing chemical waste and hazardous landfills to make the earth a safe planet for humans, plants and animals.

Environmental Chemistry and Ecotoxicology of Hazardous Heavy Metals

Environmental pollution is one of the major challenges in the modern human society. Environmental contamination and pollution by heavy metals is a threat to the environment and is of serious concern. Rapid industrialization and urbanization have caused contamination of the environment by heavy metals, and their rates of mobilization and transport in the environment have greatly accelerated since 1940s. Their natural sources in the environment include weathering of metal-containing rocks and volcanic eruptions, while principal anthropogenic sources include industrial emissions, mining, smelting, and agricultural activities like application of pesticides and phosphate fertilizers. Combustion of fossil fuels also contributes to the release of heavy metals such as cadmium (Cd) to the environment. Heavy metals are persistent in the environment, contaminate the food chains, and cause different health problems due to their toxicity. Chronic exposure to heavy metals in the environment is a real threat to living organisms.

Metal concentrations above threshold levels affect the microbiological balance of soils and can reduce their fertility. Bioaccumulation of toxic heavy metals in biota of the riverine ecosystems may have adverse effects on animals and humans. Higher levels of heavy metals in biota can have negative effects on the ecological health of aquatic animal species and may contribute to declines in their populations. Heavy metals are strong neurotoxins in fish species. The interaction of heavy metals with chemical stimuli in fish may interrupt the communication of fish with their environment. Heavy metals have been found associated with fish deformities in both natural populations and in the laboratory. Generally, such deformities have negative effects on fish populations because deformities affect their survival, growth rates, welfare, and external image. These deformities in fish can serve as excellent biomarkers of environmental heavy metal pollution. Hartl remarks that "metals, of natural or anthropogenic origin, are ubiquitous in the aquatic environment, and therefore understanding their behavior and interaction with aquatic organisms, particularly fishes, a major source of protein for human consumption, is of a great socioeconomic importance."

Metals and their Essentiality for Life

Chemically, metals are defined as "elements, which conduct electricity, have a metallic luster, are malleable and ductile, form cations, and have basic oxides". Terms usually used in relation to metals in biological and environmental studies are metal, metalloid, semimetal, light metal, heavy metal, essential metal, beneficial metal, toxic metal, abundant metal, available metal, trace metal, and micronutrient. Metals have very diverse applications and play an important role in the industry-dominated human society. Some metals have critically important physiological and biochemical functions in biological systems, and either their deficiency or excess can lead to disturbance of metabolism and therefore to various diseases. Some metals and metalloids are essential for (biological) life. They play important physiological and biochemical roles in the body as they may be part of biomolecules such as enzymes, which catalyze biochemical reactions in the body.

Heavy Metals (HMs)

According to Csuros and Csuros, a heavy metal is defined as "a metal with a density greater than 5 g/cm^3 (i.e., specific gravity greater than 5)." According to Duffus, "the term "heavy metals" is often used as a group name for metals and semimetals (metalloids) that have been associated with contamination and potential toxicity or ecotoxicity." Very recently, we have proposed a broader definition for the term, and heavy metals have been defined as "naturally occurring metals having atomic number greater than 20 and an elemental density greater than 5 g·cm^{-3}".

Essential and Non-essential HMs

Regarding their roles in biological systems, heavy metals are classified as essential and nonessential. Essential heavy metals are important for living organisms and may be required in the body in quite low concentrations. Nonessential heavy metals have no known biological role in living organisms. Examples of essential heavy metals are Mn, Fe, Cu, and Zn, while the heavy metals Cd, Pb, and Hg are toxic and are regarded as biologically nonessential. The heavy metals Mn, Fe, Co, Ni, Cu, Zn, and Mo are micronutrients or trace elements for plants. They are essential for growth and stress resistance as well as for biosynthesis and function of different biomolecules such as carbohydrates, chlorophyll, nucleic acids, growth chemicals, and secondary metabolites. Either deficiency or excess of an essential heavy metal leads to diseases or abnormal conditions. However, the lists of essential heavy metals may be different for different groups of organisms such as plants, animals, and microorganisms. It means a heavy metal may be essential for a given group of organisms but nonessential for another one. The interactions of heavy metals with different organism groups are much complex.

Environmentally Relevant most Hazardous HMs and Metalloids

Heavy metals are among the most investigated environmental pollutants. Almost any heavy metal and metalloid may be potentially toxic to biota depending upon the dose and duration of exposure. Many elements are classified into the category of heavy metals, but some are relevant in the environmental context. List of the environmentally relevant most toxic heavy metals and metalloids contains Cr, Ni, Cu, Zn, Cd, Pb, Hg, and As. Heavy metal pollutants most common in the environment are Cr, Mn, Ni, Cu, Zn, Cd, and Pb. In 2009, China has suggested four metals, i.e., Cr, Cd, Pb, Hg, and the metalloid As, as the highest priority pollutants for control in the "12th 5-year plan for comprehensive prevention and control of heavy metal pollution". Some other heavy metals are also hazardous to living organisms depending upon dose and duration of exposure. For example, Mansouri et al. have found Ag as more toxic than Hg to a freshwater fish.

Sources of Heavy Metals in the Environment

Sources of heavy metals in the environment can be both natural/geogenic/lithogenic and anthropogenic. The natural or geological sources of heavy metals in the environment include weathering of metal-bearing rocks and volcanic eruptions. The global trends of industrialization and urbanization on Earth have led to an increase in the anthropogenic share of heavy metals in the environment. The anthropogenic sources of heavy metals in the environment include mining and industrial and agricultural activities. These metals (heavy metals) are released during mining and extraction of different elements from their respective ores. Heavy metals released to the atmosphere during mining, smelting, and other industrial processes return to the land through dry and wet deposition. Discharge of wastewaters such as industrial effluents and domestic sewage add heavy metals to the environment. Application of chemical fertilizers and combustion of fossil fuels also contribute to the anthropogenic input of heavy metals in the environment. Regarding contents of heavy metals in commercial chemical fertilizers, phosphate fertilizers are particularly important.

In general, phosphate fertilizers are produced from phosphate rock (PR) by acidulation. In the acidulation of single superphosphate (SSP), sulfuric acid is used, while in acidulation of triple superphosphate (TSP), phosphoric acid is used. The final product contains all of the heavy metals present as constituents in the phosphate rock. Commercial inorganic fertilizers, particularly phosphate fertilizers, can potentially contribute to the global transport of heavy metals. Heavy metals added to agricultural soils through inorganic fertilizers may leach into groundwater and contaminate it. Phosphate fertilizers are particularly rich in toxic heavy metals. The two main pathways for transfer of toxic heavy metals from phosphate fertilizers to the human body: (i) Phosphate rock → fertilizer → soil → plant → food → human body, (ii) Phosphate rock → fertilizer → water → human body

Combustion of fossil fuels in industries, homes, and transportation is an anthropogenic source of heavy metals. Vehicle traffic is among the major anthropogenic sources of heavy metals such as Cr, Zn, Cd, and Pb. Higher concentrations of environmentally important heavy metals have been reported in soils and plants along roads in urban and metropolitan areas. Regarding anthropogenic sources of heavy metals, emissions from coal combustion and other combustion processes are very important. During coal combustion, Cd, Pb, and As are partially volatile, while Hg is fully volatile. In China, combustion of coal is one of the major sources of atmospheric emissions of hazardous trace elements. Table lists some environmentally important data for eight key hazardous trace elements (HTEs), of prime environmental concern.

Some environmentally important data for key hazardous trace elements (htes), of prime environmental concern:

Element	Simulated mean conc. (ppm)	Emissions from coal use in China in 2007 (t)
Cr	30.37	8217.8
Ni	17.44	2308.4
Cd	0.61	245.4
Pb	23.04	12547.0
Hg	0.20	305.9
As	5.78	2205.5
Se	3.66	2353.0
Sb	2.01	546.7

The anthropogenic sources of Cr include electroplating industries, leather tanneries, textile industries, and steel industries. Globally, about 50,000 t/year of Cr may be emitted from coal combustion, wood burning, and refuse incineration. Fertilizers also usually contain significant contents of Cr. Globally, about 60,000 t/year of Ni may be generated from coal combustion; its greater portion remains in the ash. The natural sources of Cd in the environment are volcanic action and weathering of rocks, whereas an anthropogenic source is nonferrous metal mining, especially processing of Pb-Zn ores. Globally, about 7,000 t/year of Cd may be emitted from coal combustion, and sewage sludge incineration is also a source of Cd. Anthropogenic increases in Cd concentrations are also caused by excessive application of chemical fertilizers. P-containing fertilizers contain Cd as a contaminant at concentrations ranging from trace quantities to 300 ppm on dry weight basis and hence may be a main source of input of this metal to agricultural systems. Pb is released to the environment from different sources including acid batteries, old plumbing systems, and lead shots used for hunting of game birds. Combustion of leaded gasoline is also a source of Pb in the environment. Although use of the tetraethyl lead as an antiknock agent in gasoline has been banned, it is still used in some developing regions of the world.

Contamination of Natural Waters, Sediments and Soils by Heavy Metals

Toxic trace metals pose an important threat to both aquatic and terrestrial ecosystems. After release from both natural and anthropogenic sources, heavy metals contaminate natural water bodies, sediments, and soils. Heavy metals released into the atmosphere in volcanic eruptions and in different industrial emissions also ultimately return to the land and cause contamination of waters and soils. Since heavy metals are persistent in the environment, they either accumulate in biota or leach down into ground waters. Contamination of biota and groundwater with potentially toxic heavy metals has important implications for human health. It is important to assess the degree of heavy metal pollution in riverine ecosystems by investigating the concentrations of these elements and their distribution. Figure shows a conceptual schematic of contamination of an aquatic (riverine) ecosystem with heavy metals. Different physicochemical and climatic factors affect the overall dynamics and biogeochemical cycling of heavy metals in the environment.

Schematic showing contamination of a river with heavy metals.

Water

It is said that water is the "life-blood of the biosphere." Since water is a universal solvent, it dissolves different organic and inorganic chemicals and environmental pollutants. Aquatic ecosystems, both freshwater and marine, are vulnerable to pollution. Contamination of water resources by heavy metals is a critical environmental issue which adversely affects plants, animals, and human health. Heavy metals are extremely toxic to aquatic organisms even at very low concentrations. These elements can cause significant histopathological alterations in tissues of aquatic organisms such as fish. Aquatic ecosystems are contaminated by heavy metals from different sources. One source of heavy metals in the aquatic ecosystems is effluents from mining operations. Other sources of water contamination with heavy metals include different industrial effluents, domestic sewage, and agricultural run-off. The release of industrial effluents without treatment into the aquatic bodies is a major source of pollution of surface and groundwater water. Pollution of water bodies with heavy metals is a worldwide problem because of the environmental persistence, bioaccumulation, and biomagnification in food chains and toxicity of these elements.

Sediments

Contamination of sediments with heavy metals is an environmentally important issue with consequences for aquatic organisms and human health. Sediments act as the main pool of metals in the aquatic environment. Their quality can indicate the status of water pollution. Sediments serve as both sink and source of heavy metals, releasing them into the water column. Continuing deposition of heavy metals in sediments can also lead to contamination of groundwater with these pollutants. The adsorption, desorption, and subsequent concentrations of heavy metals in sediments are affected by many physicochemical factors such as temperature, hydrodynamic conditions, redox state, content of organic matter and microbes, salinity, and particle size. Distribution of heavy metals in sediments is affected by chemical composition of the sediments, grain size, and content of total organic matter (TOM). An important determinant of metal bioavailability in sediments is pH. A lowering in pH increases the competition between metal ions and H^+ for binding sites in sediments and may result in dissolution of metal complexes, thereby releasing free metal ions into the water column. Higher concentrations of toxic heavy metals in riverine sediments may pose ecological risk to benthos (bottom-dwelling organisms).

Soils

Heavy metals and metalloids are released into soils from the parent material (lithogenic source) and different anthropogenic sources. Factors affecting the presence and distribution of heavy metals in soils include composition of parent rock, degree of weathering, and physical, chemical, and biological characteristics of soil and climatic conditions. Significant enrichment of heavy metals has been reported in soils receiving more input of fertilizers and Cu fungicide compared to virgin soils and soils receiving low inputs. In urban areas, soils may be contaminated with heavy metals from heavy vehicular traffic on roads. Soil samples in urban areas have elevated concentrations of Pb, out of which 45–85% is bioaccessible. The bioavailability of heavy metals in soils is very important for their fate in the environment and for their uptake in plants. Different heavy metals have different bioavailabilities in soils, and this bioavailability is dependent on metal speciation and on different physicochemical properties of soils.

Heavy Metals as Hazardous Materials

Heavy metals are considered as hazardous chemicals in the environment. Nonessential heavy metals are toxic to plants, animals, and humans at very low concentrations. Even the essential heavy metals also cause adverse health effects at high concentrations. Procedures developed for hazard identification of chemical contaminants in the aquatic environments consider three characteristic features: Persistence, bioaccumulation, and toxicity. Toxic substances which are both persistent and bioaccumulative are more hazardous.

Toxicity refers to the property of a chemical to affect survival, growth, and reproduction of an organism. Certain heavy metals have been reported to be carcinogenic, mutagenic, and teratogenic to different species depending on dose and duration of exposure. Heavy metals affect both wildlife and human health. Some species are more sensitive to heavy metals than others. The mechanisms by which heavy metals affect different organs, tissues, and systems in different organisms are very complex, and so far some of them are not fully explored. Exposure of the bivalve *Anodonta anatina* to Cd has been found to affect carbonic anhydrase (CA) in its tissues, an enzyme playing a role in osmoregulation and Ca metabolism. Cd has been considered as one of the factors likely responsible for the decline in populations of freshwater mussels due to its high toxicity, bioaccumulation potential, and transfer through food chains.

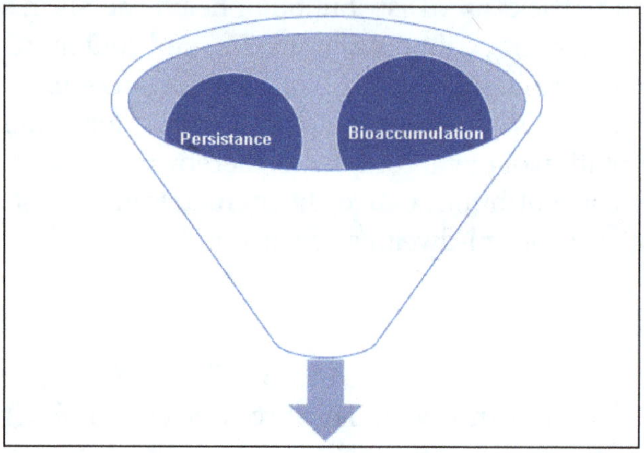

Hazard of a substance in the aquatic environment is a function of its persistence, bioaccumulation, and toxicity (PBT).

Trophic Transfer of Heavy Metals

Since heavy metals are persistent in the environment, they accumulate in living organisms and are transferred from one trophic level to another in the food chains. The extent of accumulation of heavy metals in biota depends on their rate of accumulation and their rate of elimination from the body. Thus, different heavy metals have different half lives in different species.

Heavy metals may enter the body of an organism directly from the abiotic environment, i.e., water, sediments, and soil or may enter the organism body from its food/ prey. For example, heavy metals may enter the fish body directly from water or sediments through the fish gills/skin or from the fish food/prey through its alimentary canal. The concentration of a heavy metal may increase or decrease along successive trophic levels in a food chain. The retention of heavy metals in the body of an organism depends on many factors such as the speciation of the metal concerned and the physiological mechanisms developed by the organism for the regulation, homeostasis, and detoxification of the heavy metal. Methylated forms of heavy metals such as Hg are accumulated in biota to a greater extent and therefore biomagnified in food chains due to their lipophilicity. Certain plants have the ability to thrive on metal-rich habitats and are called metallophytes. These special plants have developed special mechanisms for coping with higher heavy metal concentrations in soil and are divided into three categories, i.e., excluders, indicators, and hyperaccumulators. Certain terms are used to describe the trophic transfer of heavy metals.

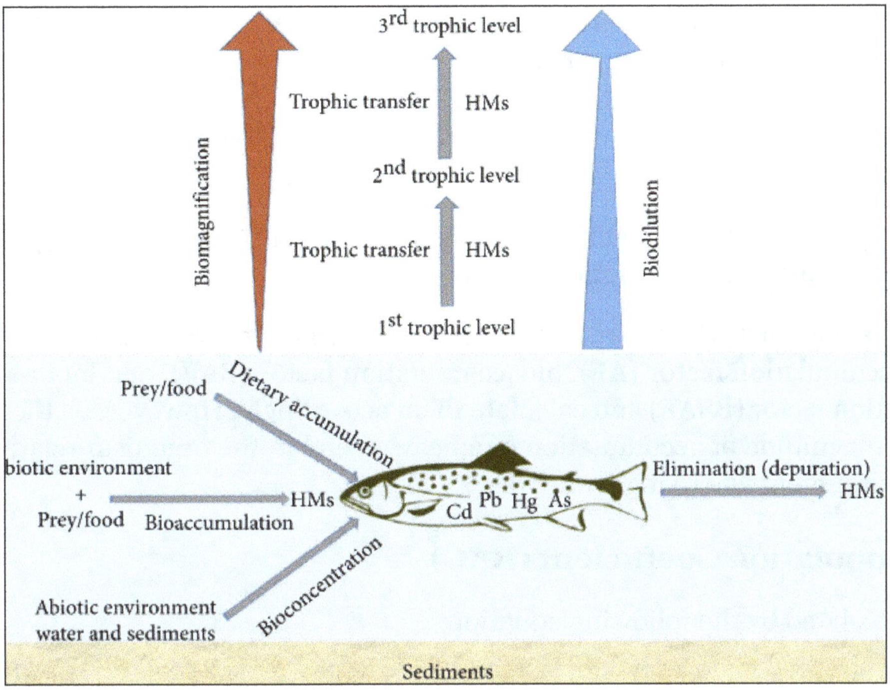

Schematic summarizing terms and concepts involved in trophodynamics of heavy metals.

Heavy Metal Transfer from Soil to Plant

The soil-to-plant transfer of heavy metals is a very important step in the trophic transfer of such metals in food chains. These metals are taken up by plants from polluted soil and subsequently transferred to herbivorous animals along the food chain. Regarding contamination of the human food chain, contamination of crops such as cereals and vegetables is a very serious issue. Consumption of cereals contaminated with toxic heavy metals may cause risk to human health. Heavy metals in higher concentration ranges have been reported in vegetables grown with wastewaters compared to those grown with groundwater. Furthermore, higher concentrations of these metals have been found in leafy vegetables compared to those in other types of vegetables such as bulbs and tubers.

Quantification of Trophic Transfer of Heavy Metals

Certain terms have been used to quantify the degree or extent of accumulation of heavy metals in biota. Some of these quantitative terms are bioconcentration factor (BCF), bioaccumulation factor (BAF), bioaccumulation coefficient (BAC), etc.

Bioconcentration Factor (BCF)

BCF is indicative of the degree of enrichment of a heavy metal in an organism relative to that in its habitat. It is defined as “the ratio of the concentration of a heavy metal in an organism tissue to its concentration in the abiotic medium (water and sediments)." It is calculated by the following equation:

$$\text{BCF} \quad \frac{\text{organism tissue}}{\text{abiotic medium}}$$

where $C_{organism}$ is the metal concentration in the organism tissue and $C_{abiotic\ medium}$ is the metal concentration in the abiotic medium.

Some researchers use the alternative terms transfer factor (TF), metal transfer factor (MTF), accumulation factor (AF), bioaccumulation factor (BAF), and biota-sediments accumulation factor (BSAF) and calculate them accordingly. However, all these indices show the magnitude of accumulation of a heavy metal in the organism relative to that in the environment where it grows/lives.

Bioaccumulation Coefficient (BAC)

BAC is calculated by the following equation:

$$\text{BAC} = \frac{C_{plant}}{C_{soul}}$$

where C_{plant} is the metal concentration in plant and is C_{soul} the metal concentration in soil.

It is evident that the values of BCF, BAF, BAC, etc. depend on the concentration of the heavy metal in the organism and in the concerned environmental medium. Since the values of these indices are inversely related to the metal concentration in the environmental medium (water, sediments, and soil), the values of these indices should be used with caution for assessment of heavy metal contamination in biota. For example, the BCF value for a heavy metal in muscles of a fish dwelling in less contaminated water may be higher than for the fish dwelling in more contaminated water simply because of the lower metal concentration in the habitat of the former fish. It has been reported that bioconcentration factor (BCF) values of seven typical heavy metals in crop grains decreased exponentially with average concentrations of the metals in soil.

Bioaccumulation of Heavy Metals in Biota

Since heavy metals are persistent in the environment, they enter from the environment to the organisms and accumulate therein. The uptake and bioaccumulation of heavy metals in biota depend on several factors. For example, the uptake of heavy metals in plants depends on bioavailability of the metal in soil, which in turn depends on several factors such as metal speciation, pH, and organic matter contents in soil. Metals which are more bioavailable in soil may be accumulated in plants more easily and thus will have more bioaccumulation potential. An assessment of bioaccumulation of heavy metals in plants may be used for an estimation of bioavailability of the metals in soil.

Such an assessment may also be used for knowing the contamination status of the environment. It has been reported that plants seem to be more sensitive to environmental changes than soils. Different plant species have been suggested as bioindicators of heavy metal pollution in the environment. Different animal species have also been suggested as bioindicators of heavy metal pollution. For example, the date mussel (*Lithophaga lithophaga*) has been suggested as a valid bioindicator of marine pollution.

Bioaccumulation of heavy metals in biota is important from environmental, ecological, and human health point of view and has important implications for wildlife and human health. Contamination of aquatic and terrestrial food chains with potentially toxic heavy metals poses a threat to the health of consumer organisms including humans. In aquatic ecosystems, organisms are simultaneously exposed to different metals, which may have additive, synergistic, or antagonistic interactions. The topic of bioaccumulation of heavy metals in biota is a very broad topic.

Bioaccumulation of Heavy Metals in Freshwater Fish

Aquatic biota is exposed to heavy metals through different routes such as water, sediments, and food. Freshwater fish are exposed to different toxic heavy metals released to freshwater bodies from different natural and anthropogenic sources. Contamination of fish by heavy metals has become an important global issue because it presents a threat to fish and poses health risks to fish consumers. Assessment of heavy metal bioaccumulation in fish species of different aquatic habitats is very important. Assessment of heavy metal levels in fish tissues is essential for aquatic ecosystem management and human consumption of fish. Fish have high levels of unsaturated fatty acids and low levels of cholesterol. They are an important source of proteins. Use of an edible fish in human diet is beneficial and therefore recommended in balanced diet. However, contamination of fish by toxic heavy metals is considered as a risk for human health and has raised concerns about their consumption especially in more sensitive groups of human population such as women, children, and people at risk of diseases from other causes.

The bioaccumulation of heavy metals in freshwater fish depends on various factors, both fish characteristics and external environmental factors. Factors related to fish include fish age, size (weight and length), feeding habits, and body physiology, while external environmental factors include concentration and bioavailability of metals in the water column, physicochemical properties of water, and other climatic factors. The degree of accumulation of heavy metals in different tissues of fish is generally different depending on the structure and function of tissues. Generally, metabolically active tissues such as gills, liver, and kidneys have higher accumulations of heavy metals than other tissues such as skin and muscles. The comparatively higher heavy metal accumulation in metabolically active tissues of fish is generally explained by the induction/occurrence of metal-binding proteins called metallothioneins (MTs) in these tissues upon exposure to heavy metals. Fish gills have been found as the target tissue for accumulation and elimination of heavy metals such as Ni. Although fish muscles are the tissue of poor heavy metal accumulation, they are important from the view point of consumption by humans. Bioaccumulation of trace metals in muscles of fish is generally species specific. Most studies on bioaccumulation of heavy metals in fish have investigated metal concentrations in fish muscles because this tissue is edible and is the most relevant regarding human health.

Bioaccumulation of toxic heavy metals in freshwater fish has important environmental, ecological, and social consequences; it has implications for humans and other fish consuming carnivorous wildlife. Waterborne heavy metals are incorporated in fish and enter the human body through the food chain and therefore have impact on human health. Furthermore, toxic heavy metals also affect fish health and well-being. It has been reported that pollution of river with heavy metals-containing wastewater induced stress in the freshwater fish *Channa punctatus*, making it weak and more vulnerable to diseases. Heavy metal pollution is regarded as one of the possible causes for population declines of freshwater fish and other aquatic species in freshwater ecosystems. It has been reported that increased contamination in River Indus, Pakistan, has caused decline in the number and diversity of freshwater fish and other aquatic species in this river.

Bioaccumulation of Heavy Metals in Rice (Oryza sativa)

Rice is a very important human food and is the staple food in Asian countries especially in South Asia and China. Contamination of rice fields with toxic heavy metals leads to bioaccumulation of these elements in the rice plant. The translocation of heavy metals from roots of the rice plant to stem, leaves, and rice grains is of human health concern. Rice crop is especially susceptible to heavy metal contamination because it needs water during most of its growth period. The trace elements, Cd, Pb, Hg and As, are ubiquitous in the environment with harmful effects on human health. Regarding their presence in rice as a public health concern, As is on top followed by Cd. Human intake of Cd has been reported to be highest through consumption of rice. Contamination of rice with toxic heavy metals is especially a health concern in developing countries.

Irrigation of agricultural soils with wastewater is a widespread practice in developing countries, which results in elevated uptake of metals in crops. Elevated levels of heavy metals in crops affect food quality and pose health risks to the consumers. Application of phosphate fertilizers rich in Cd can also lead to Cd contamination in rice fields. For the general population, consumption of rice may be a potential exposure source of toxic heavy metals, particularly Cd, Pb, and As. Long-term consumption of rice grown on polluted areas may cause potential health hazards to the consumers. Efforts are being made to minimize uptake through roots and translocation to grains of toxic heavy metals especially Cd in rice. Genetic engineering is used as an approach to achieve this goal, and some transgenic rice varieties have been developed to meet the challenge.

Bioaccumulation of Heavy Metals in Tobacco (Nicotiana Tabacum)

Bioaccumulation of toxic heavy metals in cigarette tobacco is a human health concern because tobacco leaves are used for making cigarettes. Tobacco plants naturally accumulate relatively high concentrations of heavy metals in their leaves, and the metal bioaccumulation in tobacco leaves varies with the geographical origin of the tobacco plants. Tobacco crop is grown with application of commercial inorganic fertilizers especially phosphate fertilizers, which contain considerable concentrations of some toxic heavy metals. During growth, uptake of heavy metals by tobacco roots is considerable and the same are translocated from soil to leaves. During cigarette smoking, a fraction of heavy metals is inhaled in the smoke and thus reaches the lungs of the smoker. Tobacco smoke, both mainstream and side stream, is an important source of environmental metal exposure. Passive smoking has an important contribution in the exposure of children to Pb. Heavy metals inhaled during tobacco smoking are easily absorbed in the body from the lungs and reach the blood from where they may reach other parts of the body. Higher levels of toxic heavy metals have been reported in blood of cigarette smokers compared to that of nonsmokers.

As stated by Dissanayake and Chandrajith, the application of inorganic fertilizers in agriculture has unfortunately become a "necessary evil." Since commercial chemical fertilizers are usually not purified enough during the manufacturing processes, they usually contain heavy metals as impurities. Much of the phosphate fertilizers in the world are commercially produced from phosphate rocks that contain the mineral apatite [$Ca_5(PO_4)_3OH,F,Cl$]. Due to their geological and mineralogical nature, phosphate rocks contain different environmentally hazardous elements including Cr, Cd, Pb, Hg, As, and U. Application of chemical fertilizers leads to increase in the concentrations of these potentially toxic heavy metals in agricultural soils. A high correlation has been reported between concentrations of metals, i.e., Cr, Ni, Cd, and Pb and content of phosphate in fertilizers. A study investigating concentrations of Cr, Mn, Fe, Ni, Cu, Zn, Sn, Cd, Pb, and As in tobacco products in the UK has concluded that overapplication of phosphate/nitrate fertilizers is the most likely cause of their transfer to these products. However, Bozhinova reported a limited impact of application of phosphate fertilizers on accumulation of Ni, Cu, Cd, and Pb in soil and tobacco plants. In summary, the addition of toxic heavy metals from long-term application of phosphate fertilizers to agricultural soils and their subsequent transfer to the human food chain is a matter of great concern with respect to human health, especially in case of low-quality phosphate fertilizers containing elevated levels of heavy metals.

Human Exposure to Heavy Metals

Humans are exposed to toxic heavy metals in the environment through different routes including ingestion, inhalation, and dermal absorption. People are more exposed to toxic metals in developing countries. Generally, people have no awareness and knowledge about exposure to heavy metals and its consequences for human health, especially in the developing countries. People may be exposed to heavy metals in the work place and in the environment. Human exposure to toxic chemicals in the work place is called occupational exposure while exposure to such chemicals in the general environment is called nonoccupational or environmental exposure. Workers are exposed to heavy metals in mining and industrial operations where they may inhale dust and particulate matter-containing metal particles. People extracting gold through the amalgamation process are exposed to Hg vapors. It has been reported that welders with occupational prolonged exposure to welding fumes had significantly higher levels of the heavy metals Cr, Ni, Cd, and Pb in blood than the control and showed increased oxidative stress. Cigarette smoking is also a principal source of human exposure to Cd and other toxic heavy metals present in the tobacco leaves.

Ingestion of heavy metals through food and drinking water is a major exposure source for the general human population. Industrialization, urbanization, and the rapid economic development around the globe have led to intensification in industrial and agricultural activities. Such activities may cause contamination of water, air, and soils with toxic heavy metals. Growing human foods in heavy metal-contaminated media lead to bioaccumulation of these elements in the human food chains from where these elements ultimately reach the human body.

Bioaccumulation and Biomagnification of Heavy Metals in the Human Food Chains

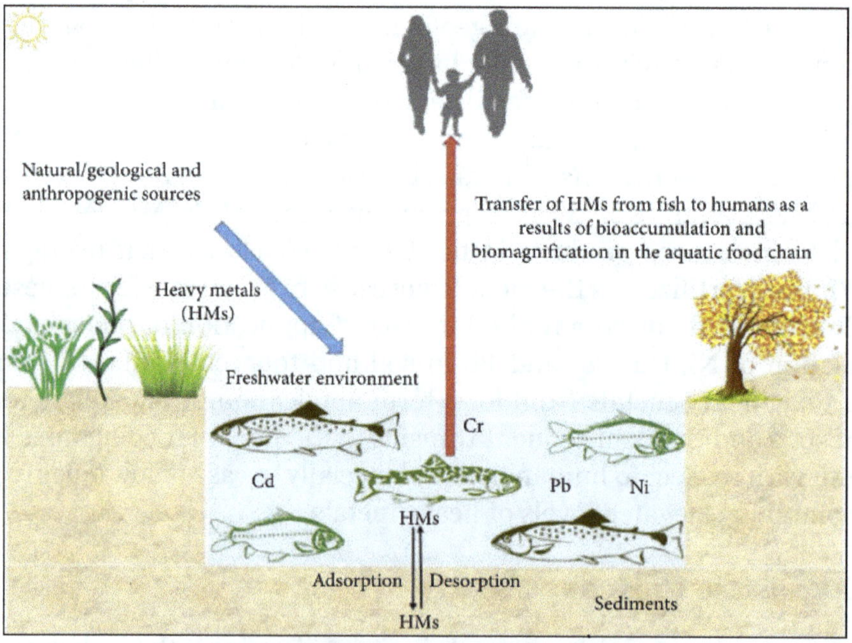

Trophic transfer of heavy metals from freshwater fish to humans in the human food chain.

Humans are omnivorous. They may be exposed to toxic heavy metals through different food items such as fish, cereals, and vegetables. Heavy metal contamination in freshwater bodies such as rivers, lakes, and streams leads to bioaccumulation of these elements in freshwater fish, while such contamination in agricultural lands leads to bioaccumulation of these elements in agricultural crops. Contamination of the human food chains with toxic heavy metals poses a threat to human health. Certain examples from the twentieth century have shown that such contamination is a serious human health issue. Minamata disease (MD) and *itai-itai* disease both in Japan were caused by consumption of Hg-contaminated fish and Cd-contaminated rice, respectively. Figure depicts the transfer of heavy metals from contaminated fish to humans.

Although biomagnification of heavy metals is a controversial issue in metal ecotoxicology, numerous studies have reported biomagnification of heavy metals in certain food chains. In case of biomagnification of these metals in food chains, organisms at higher trophic levels in the food chains are at greater risk. Higher concentrations of trace metals in organisms of higher trophic levels as a result of biomagnification can pose health risk to these organisms or to their human consumers. To protect human health from the harmful effects of toxic heavy metals, the human food chains should be constantly monitored for bioaccumulation and biomagnification of heavy metals. However, nondestructive sampling techniques and use of environmental biomarkers should be opted to avoid loss of biota due to analysis. Furthermore, in order to avoid contamination of food chains with heavy metals, untreated municipal and industrial wastewaters should not be drained into natural ecosystems such as rivers and farmlands.

Heavy Metal Toxicity

Although some heavy metals, called essential heavy metals, play important roles in biological systems, they are generally toxic to living organisms depending on dose and duration of exposure. It is a well-known fact in toxicology that "excess of everything is bad." Nonessential heavy metals (Cd, Pb, and Hg) and metalloids (As, etc.) may be toxic even at quite low concentrations. Essential heavy metals are required in trace quantities in the body but become toxic beyond certain limits or threshold concentrations. For some elements, the window of essentiality and toxicity is narrow. Heavy metals have been reported to be carcinogenic, mutagenic, and teratogenic. They cause generation of reactive oxygenic species (ROS) and thus induce oxidative stress. Oxidative stress in organisms leads to the development of various diseases and abnormal conditions. Heavy metals also act as metabolic poisons. Heavy metal toxicity is primarily due to their reaction with sulfhydryl (SH) enzyme systems and their subsequent inhibition, e.g., those enzymes involved in cellular energy production. Figure shows the reaction of a heavy metal (M) with glutathione (GSH), an important antioxidant in the body. Here, the metal replaces H atoms from SH groups on two adjacent glutathione molecules. The engagement of the two glutathione molecules in formation of strong bond with the metal deactivates them for further reactions:

$$2\,\text{glutathione} + \text{metal ion}\left(M^{2+}\right) \rightarrow M\left(\text{glutathione}\right)_2 + 2H^+$$

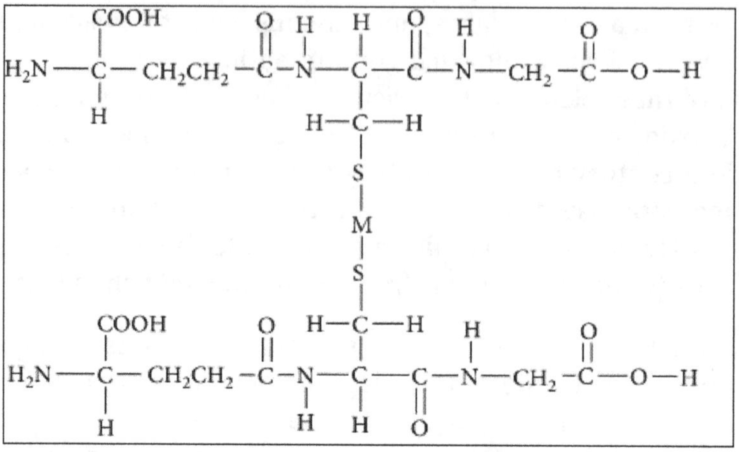

Reaction of a metal with glutathione.

Effects of Toxic Heavy Metals on Human Health

The heavy metals Cd, Pb, Hg, and As deplete the major antioxidants of cells, particularly antioxidants and enzymes having the thiol group (—SH). Such metals may increase the generation of reactive oxygen species (ROS) like hydroxyl radical ($HO^{\cdot}$), superoxide radical ($O_2^{\cdot -}$), and hydrogen peroxide (H_2O_2). Increased generation of ROS can devastate the inherent antioxidant defenses of cells and lead to a condition called "oxidative stress". Heavy metals, including Cd, Pb, and Hg, are nephrotoxic, especially in the renal cortex. The chemical form of the heavy metals is important in toxicity. Mercury toxicity largely depends on Hg speciation. Relatively higher concentrations of toxic heavy metals, i.e., Cr, Cd, and Pb, and relatively lower concentrations of the antioxidant element Se have been found in patients of cancer and diabetes compared to those in the normal subjects in Lahore city, Pakistan.

Monitoring and Analysis of Heavy Metals in the Environment

Monitoring and analysis of heavy metal concentrations in the environment are necessary for pollution assessment and control. The levels/concentrations of potentially toxic metals and metalloids should be regularly monitored in the different environmental media such as water, sediments, and soils as well as in the resident biota. Such environmental analysis will provide useful information about distribution, principal sources, and fate of these elements in the environment and their bioaccumulation in the food chains. Such analysis is also used to assess the risk posed by these elements to wildlife and human health.

Use of Bioindicators and Biomarkers for Assessment of Heavy Metal Pollution

Regarding the use of bioindicators for monitoring and assessment of heavy metal pollution, Morgan summarizes that "a more meaningful assessment of the impact of metal pollution may be obtained by measuring metal concentrations in selected species of the resident biota." Different plant and animal species have been used as biological indicators or bioindicators to assess and monitor heavy metal contamination and pollution in the environment. Different environmental biomarkers are also used to assess and monitor pollution pertaining to heavy metals in the environment.

The Nature and Scope of Studies on Heavy Metals in the Environment

Environmental studies on different aspects of heavy metals and metalloids are interdisciplinary in nature and require background knowledge in environmental chemistry, ecotoxicology, and ecology. Analysis of xenobiotics such as toxic heavy metals in food chains is an important study area and has environmental, ecological, and economic importance. It has a scope for public health. Such studies involve aquatic chemistry, which has a scope for public health as remarked by Johnston : "aquatic chemistry is a fundamental element of public health." Bioaccumulation data of toxic heavy metals in different biota such as fish and rice can be used for health risk assessment for the general human population.

Environmental Biochemistry

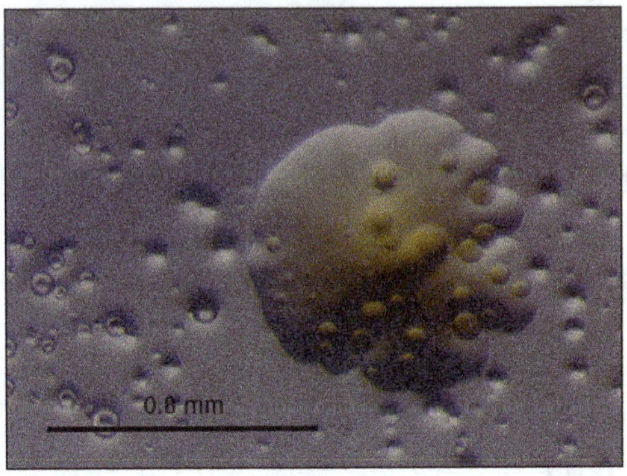

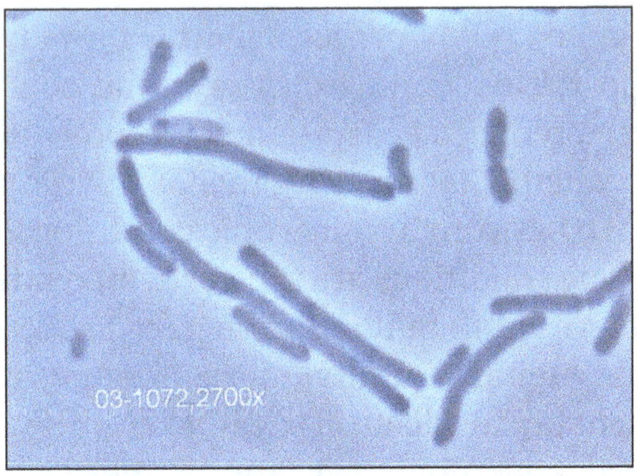

Environmental biochemistry is the application of principles of biochemistry for the protection of environment. The main themes include managing water quality and air resources, protection from radiation, to maintain industrial hygiene etc. Environmental biochemists employ living organism and their capabilities for such purposes. The environmental biochemistry is covered in studying microbial metabolism of pollutants with emphasis on elucidation of metabolites and critical metabolic reactions.

Biodegradation and microbial transformation reactions are the most important processes determining environmental fate and ultimate elimination of environmental pollutants. The environmental biochemistry group is engaged in studying microbial metabolism of pollutants with emphasis on elucidation of metabolites and critical metabolic reactions. On a molecular level, we analyze enzymes catalyzing key steps in the metabolism of pollutants with regard to genetic, kinetic, and structural criteria. The acquired insights will help us to develop strategies, tools, and methods for improvement and advancement of drinking water and waste water purification technologies, bioremediation processes, construction and selection of bacteria with specific physiological characteristics, and environmentally friendly industrial production processes for chemicals (green chemistry).

The biochemical pathways developed by many organisms for satisfying their nutrient and energy demands, often contribute to the recycling of elements in an ecosystem. The importance of such processes in the nature were early stressed which demonstrated that special groups of microorganisms normally participate in the redox reactions which occur in different habitats of the earth. The carbon, nitrogen and sulfur cycles were proposed for describing the most important processes involved in the cyclic turnover of the elements in nature.

Transformations in Nature

The complex redox equilibrium of the elements established in any habitat results from the interaction of two components, one originating in physicochemical reactions and the other from biological processes. Although the first can sometimes exclude the second, usually both are interactive and either can induce a corresponding modification in the development of the other.

As in any redox process, matter flows from an oxidized or reduced state towards the opposite, depending on the initial conditions provided by the environment. Therefore, a reduced ecosystem corresponds to a medium rich in electrons, which are easily available to microorganisms for reducing the oxidized 'energy transducer' of cells (NAD^+). Normally anaerobic conditions are representative of such media. Conversely, aerobic conditions are typical of an oxidized habitat, poor in electrons. However, habitats with different redox degrees are found in which oxygen is absent: in these conditions, organic or inorganic compounds may be used as electron acceptors. If oxygen is absent but an electron acceptor is present, the redox process that takes place in cells is called 'anaerobic respiration'. There is appreciable difference between the processes which could be expected in aerobic/anaerobic habitats or oxidized/reduced habitats.

The former are examples of biochemical pathways employed by organisms for delivering to electron acceptors the excess of electrons produced by their own metabolic pathways. Other pathways may be used by microorganisms in aerobic, anaerobic, oxidized or reduced media, for obtaining energy. In these mechanisms the intermediary molecule is NADH. Therefore the couple, $NADH/NAD^+$, represents one of the most important biological redox system that allows the organisms to obtain useful energy from the environment. A close relationship between the extracellular physicochemical characteristics of the environment and the metabolic reactions displayed by the microorganisms is observed. Highly reduced media should allow the development of microorganisms which should be able to displace the redox equilibrium towards NAD^+. This is the case in fermentations, where a suitable substrate $(cn_2o)_m$ is consumed as carbon and energy source, producing components more oxidized (CO_2) or reduced (fermentation products, $(CH_2o)_m$) depending on the particular pathways employed by the organisms. If in the later situation it is incorporated an electron acceptor, the equilibrium will tend to NAD^+ through an electron chain, inducing anaerobic respiration. For this reason, we decided to separate the redox characteristics of the medium and the availability of oxygen, as two independent variables, as shown in figure. Furthermore, if, in the above conditions and in the absence of an electron acceptor, the organisms assimilate CO_2, biosynthesis becomes unfavorable unless another electron source is present. This is the case of photoautotrophic bacteria. A similar situation may be observed under aerobic conditions when NADH is oxidized via the respiratory chain and NADPH is also consumed for carbon assimilation by using the Calvin cycle. This occurs in chemolithotrophic bacteria. From this point of view, biological matter represents an intermediate component between the completely oxidized and reduced nutrient and energy source, independently of the presence or absence of oxygen.

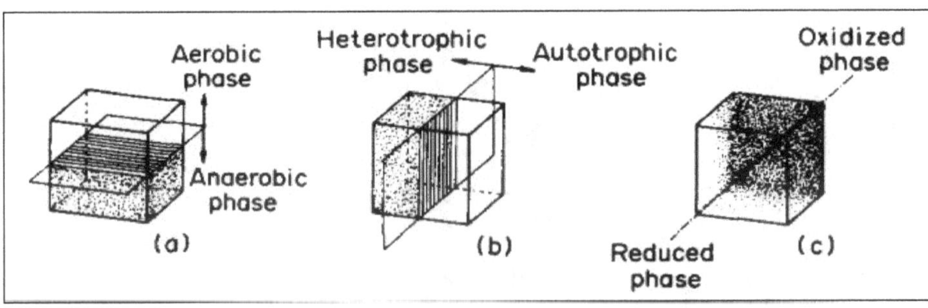

Cube division into trophic and environmental regions.

According to the importance attributed to the redox couple NADH/NAD +, we decided to incorporate this variable as a third dimension along with the two dimensional representation of the cycles of the elements (carbon, nitrogen and sulfur) but in terms of the redox degree of the environment (oxidized/reduced phases). We suggest the representation of these cycles as a cube which is divided into many regions depending on the environmental characteristics such as: aerobic and anaerobic phases, reduced and oxidized phases and two physiological regions according to the capacity for assimilation of carbon source, autotrophic and heterotrophic regions. This representation may be used to define the main metabolic feature of any organism depending on its position in the subspace of the cube. Inside the cube, aerobic and anaerobic phases should clearly be separated as should the assimilation processes of the carbon source.

Although further sophistication may be included by assigning to the cube different degrees of tolerance to oxygen or facultative autotrophic metabolism, we have not considered these. However, we have considered the degrees of reduction of the environment and matter, in the sense mentioned above, for better understanding the relationship between environmental conditions and the most important metabolisms which microorganisms may display in those habitats.

The Carbon Cycle

Including a third dimension into the well known diagram of the carbon cycle we obtained figure which shows the main processes by which carbon is oxidized or reduced according to the environmental conditions. In figure, the reduction of carbon dioxide could take place in an aerobic (upper) and an anaerobic (lower) phase. In aerobic conditions, carbon dioxide reduction takes place in plants, algae or chemolithotrophic bacteria. These processes can be found in a special region of the cube designated as I. In region II, anaerobic assimilation of carbon dioxide may be identified as the zone where photoautotrophic bacteria would be found. In both regions (I and II), the most important product of synthesis is organic matter designated as $(CH_2o)m$, which can be set on an oblique plane containing CO_2 but in a different vertical plane according to the degree of reduction of the organic compounds (ie between CH_4 and CO_2).

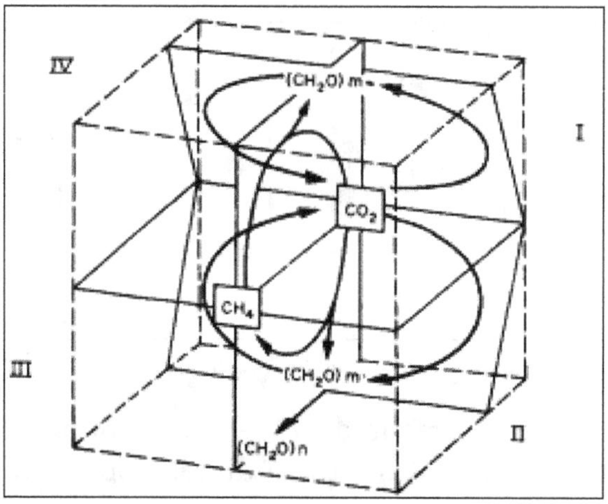

Three dimensional arrangement of the carbon cycle.

Heterotrophic fermentation and respiration, are drawn by representing the oxidative processes which occur during anaerobic (region III) or aerobic (region IV) assimilation of organic compounds respectively. In the anaerobic fermentation processes, the organic substrate induces not only the evolution of the CO_2, but also the reduction of an internal organic electron-acceptor, which constitutes the fermentation products. The latter process, was considered in figure in the pathway described as $(CH_2O)_m \rightarrow (CH_2O)_n$, where $(CH_2o)_n$ designates the reduced organic compound resulting from anaerobic fermentation.

This representation records the fact that in any biological reaction involved in microbial growth, a redox process always is taking place. It also explains the aerobic reactions and the importance of the electron acceptor in biological systems.

This sort of diagram permits the demonstration in the cube of a region corresponding to the methilotrophic and methanogenic microorganisms. If methilotrophs are not considered autotrophic bacteria, the processes carried out by this group of microorganisms may be drawn in a vertical plane between the autotrophic and the heterotrophic phases, and in the aerobic region. The processes developed by methanogenic bacteria are also included in the vertical plane, but in the anaerobic region. The latter representation clearly displays the special metabolic features which characterize those microorganisms and require them to be considered separately. If methilotrophic and methanogenic bacteria were considered as autotrophs, the vertical semiplane would be developed as oblique planes into the regions I and II respectively. Otherwise, if the latter microorganisms were considered to be heterotrophs, the oblique planes should be drawn into regions IV and III respectively. The degree of description of the main reactions involved in each process will be useful for deciding the place in the cube. We stress the versatility of the proposed diagram on including other details such as the intracellular availability of NADH or NAD^+, for reducing or oxidizing any organic compound used as energy or nutrient source.

The Nitrogen Cycle

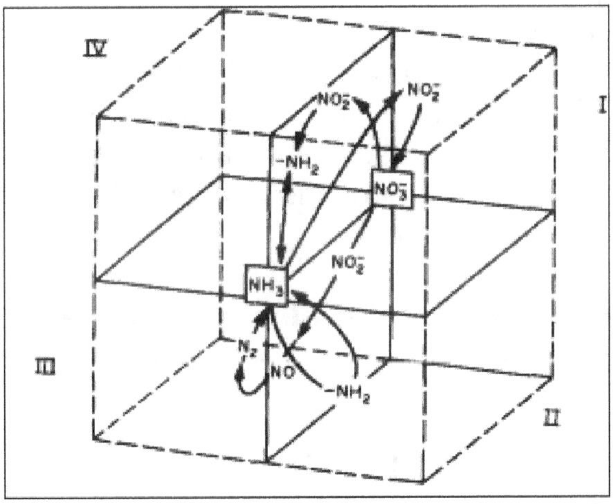

Three dimensional arrangement of the nitrogen cycle.

By using the same approach, we propose the following scheme for the nitrogen cycle. The regions described previously as I, II, III and IV, have the same meaning as in figure. The most highly oxidized (NO_3^-) and reduced (NH_3) nitrogen compounds, are placed on opposite planes, according to their degrees of oxidation. Anaerobic respiration is described by the lower oblique plane in region III, corresponding to the anaerobic heterotrophic space. Nitrate reduction or denitrification, will move the internal redox equilibrium towards NAD^+, interrupting every fermentation process. The former could be linked to the assimilative reduction of nitrate which could also be observed on the vertical plane in the anaerobic and aerobic phases.

The aerobic oxidation processes of nitrogenous compounds, carried out by chemolithotrophic bacteria, is drawn in region I of the cube. If the anaerobic reduction of nitrate induces an intracellular enrichment in NAD^+, the aerobic oxidation of NH_3 would induce a corresponding enrichment in NADH. That is the case of the metabolic reactions displayed by the chemolithotrophic nitrogen bacteria. This group of microorganisms also demands a high amount of NADH (NADPH) for CO_2 assimilation.

The Sulfur Cycle

Aerobic and anaerobic assimilative sulfate reduction can be observed in the vertical plane, between the autotrophic and heterotrophic hemispaces. Anaerobic reduction of sulfate (anaerobic respiration) is represented in region III of the cube. The processes which represent the oxidation of reduced sulfur compounds are placed in region I (aerobic) and region II (anaerobic), and both into the autotrophic zone. The line corresponding to the anaerobic oxidation of sulfur compounds cannot be represented in figure for the nitrogenous compounds, since it has not yet been described a similar process for reduced nitrogen compounds. The latter, constitutes a great difference that emerges from the cycles and are stressed by such a three dimensional (3-D) representation. The anaerobic oxidation of sulfur compounds therefore can be considered as a physiological advantage developed by photoautotrophic bacteria, to overcome the high demand of NADH (NADPH) for supplying reducing power to the CO_2 assimilation process. Other reduced sulfur intermediate compounds may be included in the semiplane which represents these sorts of processes, depending on the complexity level we wish to show. Aerobic oxidation of sulfur compounds is described in region I, similar to that for the nitrogen cycle. Chemolithotrophic sulfur bacteria are involved in these processes.

An Integrative Viewpoint for the Element Cycles

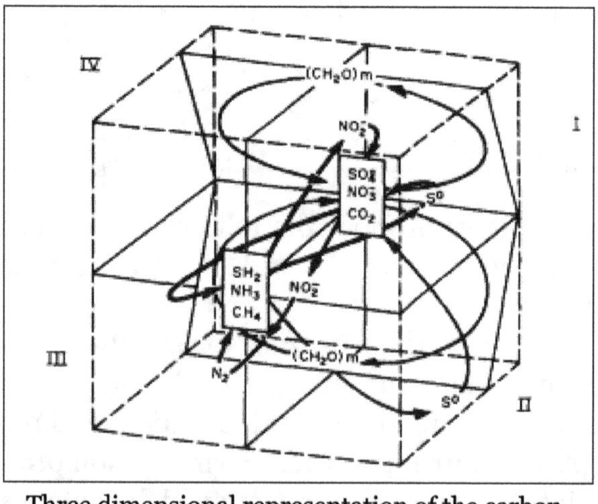

Three dimensional representation of the carbon, nitrogen and sulfur cycles into the four-region cube.

The possibility of assigning a direction and a sense to the biological processes involved in the carbon, nitrogen and sulfur cycles allowed us to consider a 3-D arrangement to represent the three cycles of the above elements. This is feasible, due to the 3-D distribution assigned to each of the biological processes represented in figures. The last feature could not be accounted for with the traditional 2-D representation of the element cycles.

Figure shows a simplified picture of the carbon, nitrogen and sulfur cycles all together. The space orientation of each biological process is according to figures, with simplification in order to clarify the scheme. It offers a comprehensive integration of the most important processes which participate in the turnover of the elements in nature.

In region I, aerobic autotrophic organisms are found. If the 'electron donor' is light, algae and plants will participate: In the dark, chemolithotrophic nitrogen and sulfur bacteria may be found.

In region II, only photoautotrophic bacteria will be found: Life Will only be possible if light is present.

In region III, the anaerobic heterotrophic way of life takes place. Fermentation and anaerobic respiration (nitrate and sulfate reduction), contribute to the enrichment in organic matter to those anaerobic ecosystems. In these heterotrophic habitats, life does not depend on light, although some facultative phototrophic bacteria and other higher organisms may develop.

Region IV, represents the space devoted to aerobic heterotrophic processes in which only the oxidation of organic substrates occurs. Preformed organic compounds, from anaerobic or aerobic processes, produced by autotrophic organisms, are used for providing carbon and energy to other aerobic or anaerobic obligate heterotrophic organisms.

Many other conclusions can be inferred from this 3-D representation of the element cycles. The conspicuous population of microorganisms which can be detected in each region does not exclude the possibility of finding some of them displaying the capacity to develop in two or three regions. When such microorganisms are detected, none of them can display processes situated in different regions at the same time. For example, if E coli can be placed according to its metabolic features in regions III and IV, it means that one condition must exclude the other. When E coli develops an aerobic metabolism, it cannot ferment or reduce nitrate simultaneously. This feature may be a useful example for introducing the students to the notion of metabolic regulation. Another interesting example is the recently reported novel energy metabolism involving fermentation of inorganic sulfur compounds. Thus Desulfovibrio sulfodismutans can oxidize sulfide and other reduced intermediate compounds anaerobically to sulfate while a reduction to sulfide is also taken place. When this sort of metabolism is displayed, D sulfodismutans can use only acetate as the sole carbon source, but when the carbon source is other than acetate (lactate, ethanol, propanol, etc) D sulfodismutans grows only by dissimilatory sulfate reduction, producing sulfide from sulfate. In both cases the metabolism of this microorganism may occupy region III in the cube, but depending on which of them is taking place, the line representing the sulfate reduction will change to a double sense one in order to describe the anaerobic heterotrophic oxidation of sulfur compounds. The variability depends on the microorganisms, but each subspace in the cube could constitute a closed region. This point of view of the biochemical processes which take place in nature, will be very helpful on teaching environmental biochemistry.

References

- How-many-types-of-environment-are-there, biology: byjus.com, Retrieved 27 February, 2020
- Environmental-chemistry, earth-and-planetary-sciences: sciencedirect.com, Retrieved 2 March, 2020
- Environmental-biochemistry, main-focus, umik, department: eawag.ch, Retrieved 7 May, 2020
- Importance-of-environmental-chemistry: incaweb.org, Retrieved 18 February, 2020

2

Atmospheric and Water Chemistry

The branch of atmospheric science that is concerned with the chemical processes occurring within the Earth's atmosphere is called atmospheric chemistry. Water chemistry studies chemical components and properties of water. This is an introductory chapter which will briefly introduce about atmospheric and water chemistry.

Atmospheric Chemistry

Atmospheric chemistry is the branch of atmospheric science focused on chemical processes within the Earth's atmosphere, including photochemistry of gas compounds; the formation and properties of airborne aerosol particles; gas–particle interactions; cloud processes; emission, transport, and dispersion of chemical tracers; and biogeochemical cycles (i.e., interaction with the underlying surface). The motivations to understand these processes include the relevance to climate forcing, implications for human health, and reciprocal interactions with (i.e., exchange of mass with) the biosphere and hence terrestrial and marine ecosystem health.

Atmospheric chemistry studies were originally concerned with determining the major gases in the Earth's atmosphere. In the latter half of the last century, as air pollution became an increasing problem in many large cities, attention turned to identifying the sources, properties, and effects of the myriad of chemical species that exist in the natural and polluted atmosphere. Acid deposition, which became recognized as a widespread problem in the 1970s, led to the realization that chemical species emitted into the atmosphere can be transported over large distances and undergo significant transformations as they move along their trajectories. The identification in 1985 of significant depletion of ozone in the Antarctic stratosphere focused attention on stratospheric chemistry and the susceptibility of the stratosphere to modification. More recently, studies of the effects of trace chemical constituents in the atmosphere on the climate of the Earth have moved to center stage.

Atmospheric chemistry consists of complex interactions of elementary reactions. Some of the processes that follows on unimolecular and termolecular reactions also involve several steps, which is why the subject of multistep reactions is introduced at this stage.

Consecutive and *parallel* steps involve reactive *intermediates* in competitive processes. Reaction intermediates of particular interest include atoms, radicals, ions, and excited species. Most of these intermediates are highly reactive, and, with one or two exceptions, cannot be 'stored' in a laboratory for long periods because they are lost on the walls of the containing vessel, or react with each other. Such intermediates are not necessarily *unstable*, and chemical lifetimes of isolated atoms or radicals in the absence of surfaces can be virtually infinite. Many excited-state species are unstable, since they may possess enough internal energy to fragment, and they may also be able to lose their energy by emission of radiation. An excited species that cannot undergo loss by an allowed radiative transition is said to be *metastable*.

Multistep reaction schemes are interpreted kinetically by writing down the differential equations, for all the species of interest, including the intermediates. Solution of these equations then allows prediction of the concentration–time variation of each of the species. Unfortunately, analytical solution of the many simultaneous differential equations is rarely possible. Numerical solution has become a widely used alternative since the advent of high-speed computers and the development of good techniques for dealing with differential equations. For some highly reactive intermediates, the *Stationary-State Hypothesis* (SSH) (often alternatively called the *Steady-State Hypothesis*) provides a simplification that will permit algebraic solution of the kinetic equations. Consider an intermediate X that is created in a process whose rate is constant, and whose loss rate increases with increased [X]. After the reaction is started, [X] will increase until the rate of loss is equal to the rate of formation. A steady state for [X] has been reached, and d[X]/d*t* → 0. To illustrate the stationary-state method, consider the pair of reactions:

$$A + B \rightarrow X \quad k_f \qquad \text{formation}$$

$$X + C \rightarrow D \quad k_t \qquad \text{loss}$$

For simplicity, let [B] and [C] be in great excess, so that we may write pseudo-first-order rate coefficients $k_f' = k_f[B]$ and $k_t' = k_t[C]$. The kinetic equation that describes reactions above is:

$$\frac{d[X]}{dt} = k_f'[A] - k_t'[X]$$

If X is in a stationary state, then we set the differential equal to zero:

$$[X]_{SS} = \frac{k_f'[A]}{k_t'}$$

where $[X]_{ss}$ indicates a steady-state concentration of X. The problem is to know if the concentration of X calculated using the SSH bears any relationship to actual concentrations. So long as [A] and [C] are independent of time, $\frac{d[X]}{dt} = k_f'[A] - k_t'[X]$ can be integrated to yield:

$$[X] = \frac{k_f'[A]}{k_t'}\left[1 - \exp(-k_t')\right]$$

where *t* is the time for which the system has been reacting. This expression for [X] approaches the steady-state expression so long as $k_t't \gg 1$ the error in applying the SSH being less than 1% for $k_t't > 4.6$. The SSH can thus be applied so long as [A] and [C] remain constant over a period long enough for this inequality to be reached. It is evident that the circumstances under which the SSH is most likely to be valid are thus those where $k_t't$ is large: That is, if the species X is highly reactive. One example of a species at steady state in the atmosphere is the highly reactive electronically excited state of atomic oxygen, $O(^1D)$, throughout the troposphere, stratosphere, and probably the mesosphere. Ground-state atomic oxygen, $O(^3P)$, however, cannot generally be treated in the atmosphere by steady-state methods because of its relatively small reactivity.

Atmospheric Pollution

Air pollution occurs when harmful or excessive quantities of substances are introduced into Earth's atmosphere. Sources of air pollution include gases (such as ammonia, carbon monoxide, sulfur dioxide, nitrous oxides, methane and chlorofluorocarbons), particulates (both organic and inorganic), and biological molecules. It may cause diseases, allergies and even death to humans; it may also cause harm to other living organisms such as animals and food crops, and may damage the natural or built environment. Both human activity and natural processes can generate air pollution.

Air pollution from a coking oven.

Air pollution is a significant risk factor for a number of pollution-related diseases, including respiratory infections, heart disease, COPD, stroke and lung cancer. The human health effects of poor air quality are far reaching, but principally affect the body's respiratory system and the cardiovascular system. Individual reactions to air pollutants depend on the type of pollutant a person is exposed to, the degree of exposure, and the individual's health status and genetics. Indoor air pollution and poor urban air quality are listed as two of the world's worst toxic pollution problems in the 2008 Blacksmith Institute World's Worst Polluted Places report. Outdoor air pollution alone causes 2.1 to 4.21 million deaths annually. Overall, air pollution causes the deaths of around 7 million people worldwide each year, and is the world's largest single environmental health risk.

Productivity losses and degraded quality of life caused by air pollution are estimated to cost the world economy $5 trillion per year. Various pollution control technologies and strategies are available to reduce air pollution.

Pollutants

Before flue-gas desulfurization was installed, the emissions from this power plant in New Mexico contained excessive amounts of sulfur dioxide.

An air pollutant is a material in the air that can have adverse effects on humans and the ecosystem. The substance can be solid particles, liquid droplets, or gases. A pollutant can be of natural origin or man-made. Pollutants are classified as primary or secondary. Primary pollutants are usually produced by processes such as ash from a volcanic eruption. Other examples include carbon monoxide gas from motor vehicle exhausts or sulfur dioxide released from factories. Secondary pollutants are not emitted directly. Rather, they form in the air when primary pollutants react or interact. Ground level ozone is a prominent example of a secondary pollutant. Some pollutants may be both primary and secondary: they are both emitted directly and formed from other primary pollutants.

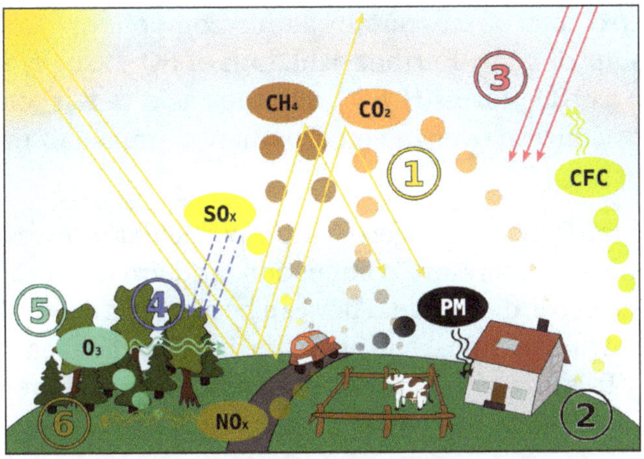

Schematic drawing, causes and effects of air pollution: (1) greenhouse effect, (2) particulate contamination, (3) increased UV radiation, (4) acid rain, (5) increased ground-level ozone concentration, (6) increased levels of nitrogen oxides.

Thermal oxidisers are air pollution abatement options for hazardous air pollutants (HAPs), volatile organic compounds (VOCs), and odorous emissions.

Pollutants emitted into the atmosphere by human activity include:

- Carbon dioxide (CO_2): Because of its role as a greenhouse gas it has been described as “the leading pollutant” and “the worst climate pollutant”. Carbon dioxide is a natural component of the atmosphere, essential for plant life and given off by the human respiratory system. This question of terminology has practical effects, for example as determining whether the U.S. Clean Air Act is deemed to regulate CO_2 emissions. CO_2 currently forms about 410 parts per million (ppm) of earth’s atmosphere, compared to about 280 ppm in pre-industrial times, and billions of metric tons of CO_2 are emitted annually by burning of fossil fuels. CO_2 increase in earth’s atmosphere has been accelerating.
- Sulfur oxides (SO_x): Particularly sulfur dioxide, a chemical compound with the formula SO_2. SO_2 is produced by volcanoes and in various industrial processes.

Coal and petroleum often contain sulfur compounds, and their combustion generates sulfur dioxide. Further oxidation of SO_2, usually in the presence of a catalyst such as NO_2, forms H_2SO_4, and thus acid rain is formed. This is one of the causes for concern over the environmental impact of the use of these fuels as power sources.

- Nitrogen oxides (NO_x): Nitrogen oxides, particularly nitrogen dioxide, are expelled from high temperature combustion, and are also produced during thunderstorms by electric discharge. They can be seen as a brown haze dome above or a plume downwind of cities. Nitrogen dioxide is a chemical compound with the formula NO_2. It is one of several nitrogen oxides. One of the most prominent air pollutants, this reddish-brown toxic gas has a characteristic sharp, biting odor.
- Carbon monoxide (CO): CO is a colorless, odorless, toxic gas. It is a product of combustion of fuel such as natural gas, coal or wood. Vehicular exhaust contributes to the majority of carbon monoxide let into our atmosphere. It creates a smog type formation in the air that has been linked to many lung diseases and disruptions to the natural environment and animals.
- Volatile organic compounds (VOC): VOCs are a well-known outdoor air pollutant. They are categorized as either methane (CH_4) or non-methane (NMVOCs). Methane is an extremely efficient greenhouse gas which contributes to enhanced global warming. Other hydrocarbon VOCs are also significant greenhouse gases because of their role in creating ozone and prolonging the life of methane in the atmosphere. This effect varies depending on local air quality. The aromatic NMVOCs benzene, toluene and xylene are suspected carcinogens and may lead to leukemia with prolonged exposure. 1,3-butadiene is another dangerous compound often associated with industrial use.
- Particulate matter/particles, alternatively referred to as particulate matter (PM), atmospheric particulate matter, or fine particles, are tiny particles of solid or liquid suspended in a gas. In contrast, aerosol refers to combined particles and gas. Some particulates occur naturally, originating from volcanoes, dust storms, forest and grassland fires, living vegetation, and sea spray. Human activities, such as the burning of fossil fuels in vehicles, power plants and various industrial processes also generate significant amounts of aerosols. Averaged worldwide, anthropogenic aerosols—those made by human activities—currently account for approximately 10 percent of our atmosphere. Increased levels of fine particles in the air are linked to health hazards such as heart disease, altered lung function and lung cancer. Particulates are related to respiratory infections and can be particularly harmful to those already suffering from conditions like asthma.
- Persistent free radicals connected to airborne fine particles are linked to cardiopulmonary disease.
- Toxic metals, such as lead and mercury, especially their compounds.

- Chlorofluorocarbons (CFCs): Harmful to the ozone layer; emitted from products are currently banned from use. These are gases which are released from air conditioners, refrigerators, aerosol sprays, etc. On release into the air, CFCs rise to the stratosphere. Here they come in contact with other gases and damage the ozone layer. This allows harmful ultraviolet rays to reach the earth's surface. This can lead to skin cancer, eye disease and can even cause damage to plants.
- Ammonia: Emitted mainly by agricultural waste. Ammonia is a compound with the formula NH_3. It is normally encountered as a gas with a characteristic pungent odor. Ammonia contributes significantly to the nutritional needs of terrestrial organisms by serving as a precursor to foodstuffs and fertilizers. Ammonia, either directly or indirectly, is also a building block for the synthesis of many pharmaceuticals. Although in wide use, ammonia is both caustic and hazardous. In the atmosphere, ammonia reacts with oxides of nitrogen and sulfur to form secondary particles.
- Odors: Such as from garbage, sewage, and industrial processes
- Radioactive pollutants: Produced by nuclear explosions, nuclear events, war explosives, and natural processes such as the radioactive decay of radon.

Secondary pollutants include:

- Particulates created from gaseous primary pollutants and compounds in photochemical smog. Smog is a kind of air pollution. Classic smog results from large amounts of coal burning in an area caused by a mixture of smoke and sulfur dioxide. Modern smog does not usually come from coal but from vehicular and industrial emissions that are acted on in the atmosphere by ultraviolet light from the sun to form secondary pollutants that also combine with the primary emissions to form photochemical smog.
- Ground level ozone (O_3) formed from NO_x and VOCs. Ozone (O_3) is a key constituent of the troposphere. It is also an important constituent of certain regions of the stratosphere commonly known as the Ozone layer. Photochemical and chemical reactions involving it drive many of the chemical processes that occur in the atmosphere by day and by night. At abnormally high concentrations brought about by human activities (largely the combustion of fossil fuel), it is a pollutant and a constituent of smog.
- Peroxyacetyl nitrate ($C_2H_3NO_5$) – similarly formed from NO_x and VOCs.

Minor air pollutants include:

- A large number of minor hazardous air pollutants. Some of these are regulated in USA under the Clean Air Act and in Europe under the Air Framework Directive.
- A variety of persistent organic pollutants, which can attach to particulates.

Persistent organic pollutants (POPs) are organic compounds that are resistant to environmental degradation through chemical, biological, and photolytic processes. Because of this, they have been observed to persist in the environment, to be capable of long-range transport, bioaccumulate in human and animal tissue, biomagnify in food chains, and to have potentially significant impacts on human health and the environment.

Sources of Air Pollutants

Mean acidifying emissions (air pollution) of different foods per 100g of protein	
Food Types	Acidifying Emissions (g SO_2eq per 100g protein)
Beef	343.6
Cheese	165.5
Pork	142.7
Lamb and Mutton	139.0
Farmed Crustaceans	133.1
Poultry	102.4
Farmed Fish	65.9
Eggs	53.7
Groundnuts	22.6
Peas	8.5
Tofu	6.7

Anthropogenic (Man-made) Sources

Controlled burning of a field outside of Statesboro, Georgia in preparation for spring planting.

Smoking of fish over an open fire.

These are mostly related to the burning of fuel:

- Stationary sources include smoke stacks of fossil fuel power stations (see for example environmental impact of the coal industry), manufacturing facilities (factories) and waste incinerators, as well as furnaces and other types of fuel-burning heating devices. In developing and poor countries, traditional biomass burning is the major source of air pollutants; traditional biomass includes wood, crop waste and dung.

- Mobile sources include motor vehicles, marine vessels, and aircraft.
- Controlled burn practices in agriculture and forest management. Controlled or prescribed burning is a technique sometimes used in forest management, farming, prairie restoration or greenhouse gas abatement. Fire is a natural part of both forest and grassland ecology and controlled fire can be a tool for foresters. Controlled burning stimulates the germination of some desirable forest trees, thus renewing the forest.

There are also sources from processes other than combustion:

- Fumes from paint, hair spray, varnish, aerosol sprays and other solvents. These can be substantial; emissions from these sources was estimated to account for almost half of pollution from volatile organic compounds in the Los Angeles basin in the 2010s.
- Waste deposition in landfills, which generate methane. Methane is highly flammable and may form explosive mixtures with air. Methane is also an asphyxiant and may displace oxygen in an enclosed space. Asphyxia or suffocation may result if the oxygen concentration is reduced to below 19.5% by displacement.
- Military resources, such as nuclear weapons, toxic gases, germ warfare and rocketry.
- Fertilized farmland may be a major source of nitrogen oxides.

Natural Sources

Dust storm approaching Stratford, Texas.

- Dust from natural sources, usually large areas of land with little vegetation or no vegetation.
- Methane, emitted by the digestion of food by animals, for example cattle.
- Radon gas from radioactive decay within the Earth's crust. Radon is a colorless, odorless, naturally occurring, radioactive noble gas that is formed from the decay of radium. It is considered to be a health hazard. Radon gas from natural sources

can accumulate in buildings, especially in confined areas such as the basement and it is the second most frequent cause of lung cancer, after cigarette smoking.

- Smoke and carbon monoxide from wildfires. During periods of actives wildfires, smoke from uncontrolled biomass combustion can make up almost 75% of all air pollution by concentration.
- Vegetation, in some regions, emits environmentally significant amounts of Volatile organic compounds (VOCs) on warmer days. These VOCs react with primary anthropogenic pollutants—specifically, NO_x, SO_2, and anthropogenic organic carbon compounds — to produce a seasonal haze of secondary pollutants. Black gum, poplar, oak and willow are some examples of vegetation that can produce abundant VOCs. The VOC production from these species result in ozone levels up to eight times higher than the low-impact tree species.
- Volcanic activity, which produces sulfur, chlorine, and ash particulates.

Emission Factors

Beijing air on a 2005-day after rain (left) and a smoggy day (right).

Air pollutant emission factors are reported representative values that attempt to relate the quantity of a pollutant released to the ambient air with an activity associated with the release of that pollutant. These factors are usually expressed as the weight of pollutant divided by a unit weight, volume, distance, or duration of the activity emitting the pollutant (e.g., kilograms of particulate emitted per tonne of coal burned). Such factors facilitate estimation of emissions from various sources of air pollution. In most cases, these factors are simply averages of all available data of acceptable quality, and are generally assumed to be representative of long-term averages.

There are 12 compounds in the list of persistent organic pollutants. Dioxins and furans are two of them and intentionally created by combustion of organics, like open burning of plastics. These compounds are also endocrine disruptors and can mutate the human genes.

The United States Environmental Protection Agency has published a compilation of air pollutant emission factors for a wide range of industrial sources. The United Kingdom,

Australia, Canada and many other countries have published similar compilations, as well as the European Environment Agency.

Exposure

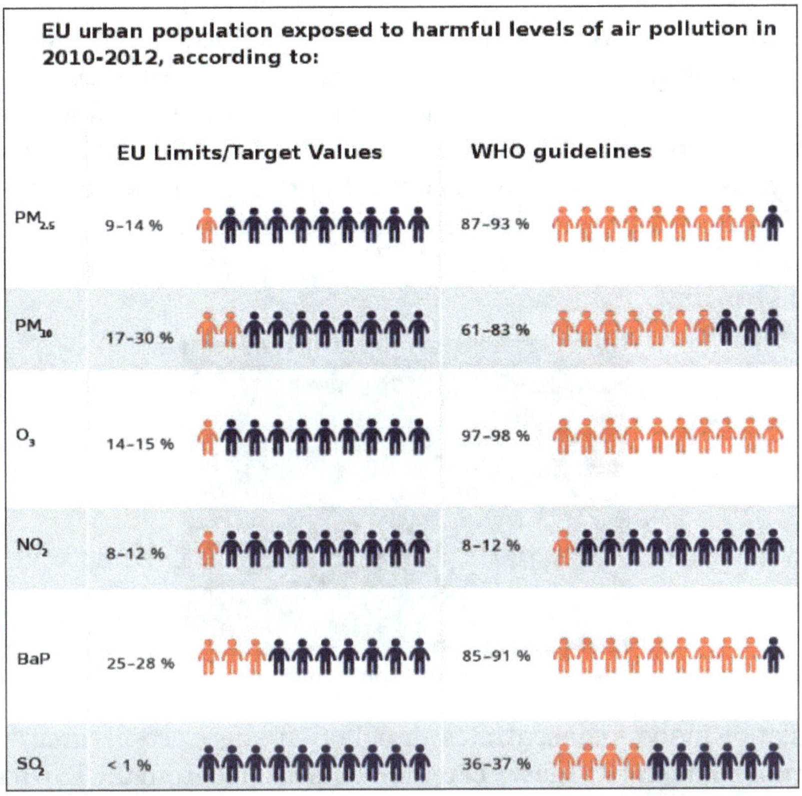

Up to 30 % of Europeans living in cities are exposed to air pollutant levels exceeding EU air quality standards. And around 98 % of Europeans living in cities are exposed to levels of air pollutants deemed damaging to health by the World Health Organization's more stringent guidelines.

Air pollution risk is a function of the hazard of the pollutant and the exposure to that pollutant. Air pollution exposure can be expressed for an individual, for certain groups (e.g. neighborhoods or children living in a country), or for entire populations. For example, one may want to calculate the exposure to a hazardous air pollutant for a geographic area, which includes the various microenvironments and age groups. This can be calculated as an inhalation exposure. This would account for daily exposure in various settings (e.g. different indoor micro-environments and outdoor locations). The exposure needs to include different age and other demographic groups, especially infants, children, pregnant women and other sensitive subpopulations. The exposure to an air pollutant must integrate the concentrations of the air pollutant with respect to the time spent in each setting and the respective inhalation rates for each subgroup for each specific time that the subgroup is in the setting and engaged in particular activities (playing, cooking, reading, working, spending time in traffic, etc.). For example, a small child's inhalation rate will be less than that of an adult. A child engaged in vigorous exercise will have a higher respiration rate than the same child in a sedentary activity. The daily exposure, then, needs to reflect the time spent in each micro-environmental setting and the type of activities in these settings. The air pollutant concentration in each microactivity/microenvironmental setting is summed to indicate the exposure. For some pollutants such as black carbon, traffic related exposures may dominate total exposure despite short exposure times since high concentrations coincide with proximity to major roads or participation to (motorized) traffic. A large portion of total daily exposure occurs as short peaks of high concentrations, but it remains unclear how to define peaks and determine their frequency and health impact.

Indoor Air Quality

Air quality monitoring.

A lack of ventilation indoors concentrates air pollution where people often spend the majority of their time. Radon (Rn) gas, a carcinogen, is exuded from the Earth in certain locations and trapped inside houses. Building materials including carpeting and plywood emit formaldehyde (H_2CO) gas. Paint and solvents give off volatile organic compounds (VOCs) as they dry. Lead paint can degenerate into dust and be inhaled. Intentional air pollution is introduced with the use of air fresheners, incense, and other scented items. Controlled wood fires in stoves and fireplaces can add significant amounts of smoke particulates into the air, inside and out. Indoor pollution fatalities may be caused by using pesticides and other chemical sprays indoors without proper ventilation.

Carbon monoxide poisoning and fatalities are often caused by faulty vents and chimneys, or by the burning of charcoal indoors or in a confined space, such as a tent. Chronic carbon monoxide poisoning can result even from poorly-adjusted pilot lights. Traps are built into all domestic plumbing to keep sewer gas and hydrogen sulfide, out of interiors. Clothing emits tetrachloroethylene, or other dry cleaning fluids, for days after dry cleaning.

Though its use has now been banned in many countries, the extensive use of asbestos in industrial and domestic environments in the past has left a potentially very dangerous material in many localities. Asbestosis is a chronic inflammatory medical condition affecting the tissue of the lungs. It occurs after long-term, heavy exposure to asbestos from asbestos-containing materials in structures. Sufferers have severe dyspnea (shortness of breath) and are at an increased risk regarding several different types of lung cancer. As clear explanations are not always stressed in non-technical literature, care should be taken to distinguish between several forms of relevant diseases. According to the World Health Organization (WHO), these may defined as; asbestosis, *lung cancer*, and *Peritoneal Mesothelioma* (generally a very rare form of cancer, when more widespread it is almost always associated with prolonged exposure to asbestos).

Biological sources of air pollution are also found indoors, as gases and airborne particulates. Pets produce dander, people produce dust from minute skin flakes and decomposed hair, dust mites in bedding, carpeting and furniture produce enzymes and micrometre-sized fecal droppings, inhabitants emit methane, mold forms on walls and generates mycotoxins and spores, air conditioning systems can incubate Legionnaires' disease and mold, and houseplants, soil and surrounding gardens can produce pollen, dust, and mold. Indoors, the lack of air circulation allows these airborne pollutants to accumulate more than they would otherwise occur in nature.

Health Effects

In 2012, air pollution caused premature deaths on average of 1 year in Europe, and was a significant risk factor for a number of pollution-related diseases, including respiratory infections, heart disease, COPD, stroke and lung cancer. The health effects caused by air pollution may include difficulty in breathing, wheezing, coughing, asthma and worsening of existing respiratory and cardiac conditions. These effects can result in increased medication use, increased doctor or emergency department visits, more hospital admissions and premature death. The human health effects of poor air quality are far reaching, but principally affect the body's respiratory system and the cardiovascular system. Individual reactions to air pollutants depend on the type of pollutant a person is exposed to, the degree of exposure, and the individual's health status and genetics. The most common sources of air pollution include particulates, ozone, nitrogen dioxide, and sulfur dioxide. Children aged less than five years that live in developing countries are the most vulnerable population in terms of total deaths attributable to indoor and outdoor air pollution.

Mortality

The World Health Organization estimated in 2014 that every year air pollution causes the premature death of some 7 million people worldwide. Studies published in March 2019 indicated that the number may be around 8.8 million.

India has the highest death rate due to air pollution. India also has more deaths from asthma than any other nation according to the World Health Organization. In December 2013 air pollution was estimated to kill 500,000 people in China each year. There is a positive correlation between pneumonia-related deaths and air pollution from motor vehicle emissions.

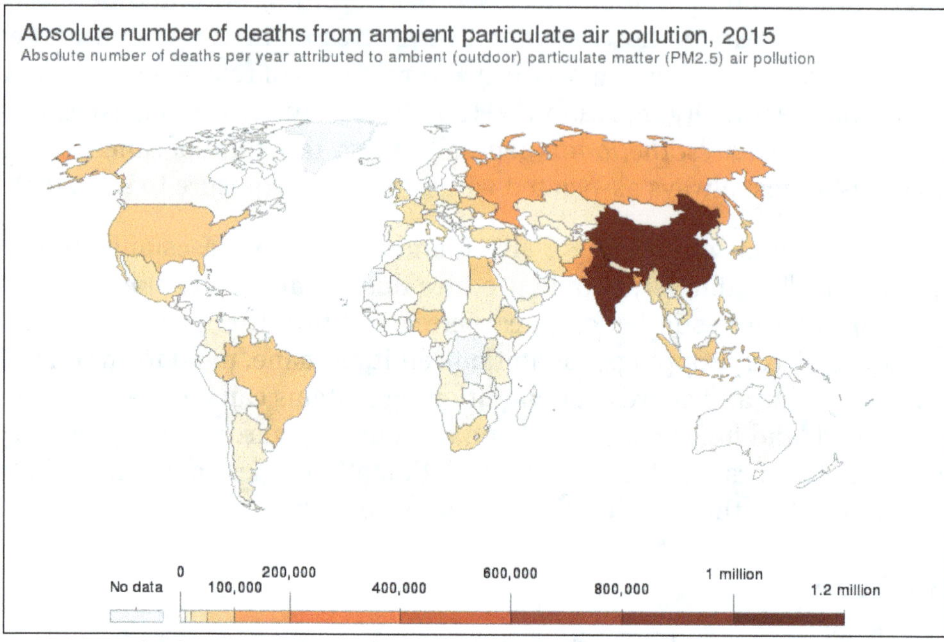

Absolute number of deaths from ambient particulate air pollution.

Annual premature European deaths caused by air pollution are estimated at 430,000-800,000 An important cause of these deaths is nitrogen dioxide and other nitrogen oxides (NOx) emitted by road vehicles. In a 2015 consultation document the UK government disclosed that nitrogen dioxide is responsible for 23,500 premature UK deaths per annum. Across the European Union, air pollution is estimated to reduce life expectancy by almost nine months. Causes of deaths include strokes, heart disease, COPD, lung cancer, and lung infections.

Urban outdoor air pollution is estimated to cause 1.3 million deaths worldwide per year. Children are particularly at risk due to the immaturity of their respiratory organ systems.

The US EPA estimated in 2004 that a proposed set of changes in diesel engine technology (*Tier 2*) could result in 12,000 fewer *premature mortalities*, 15,000 fewer heart attacks, 6,000 fewer emergency department visits by children with asthma, and 8,900 fewer respiratory-related hospital admissions each year in the United States.

The US EPA has estimated that limiting ground-level ozone concentration to 65 parts per billion, would avert 1,700 to 5,100 premature deaths nationwide in 2020 compared with the 75-ppb standard. The agency projected the more protective standard would also prevent an additional 26,000 cases of aggravated asthma, and more than a million cases of missed work or school. Following this assessment, the EPA acted to protect public health by lowering the National Ambient Air Quality Standards (NAAQS) for ground-level ozone to 70 parts per billion (ppb).

A new economic study of the health impacts and associated costs of air pollution in the Los Angeles Basin and San Joaquin Valley of Southern California shows that more than 3,800 people die prematurely (approximately 14 years earlier than normal) each year because air pollution levels violate federal standards. The number of annual premature deaths is considerably higher than the fatalities related to auto collisions in the same area, which average fewer than 2,000 per year.

Diesel exhaust (DE) is a major contributor to combustion-derived particulate matter air pollution. In several human experimental studies, using a well-validated exposure chamber setup, DE has been linked to acute vascular dysfunction and increased thrombus formation.

The mechanisms linking air pollution to increased cardiovascular mortality are uncertain, but probably include pulmonary and systemic inflammation.

Cardiovascular Disease

A 2007 review of evidence found that, ambient air pollution exposure is a risk factor correlating with increased total mortality from cardiovascular events (range: 12% to 14% per 10 microg/m^3 increase).

Air pollution is also emerging as a risk factor for stroke, particularly in developing countries where pollutant levels are highest. A 2007 study found that in women, air pollution is not associated with hemorrhagic but with ischemic stroke. Air pollution was also found to be associated with increased incidence and mortality from coronary stroke in a cohort study in 2011. Associations are believed to be causal and effects may be mediated by vasoconstriction, low-grade inflammation and atherosclerosis. Other mechanisms such as autonomic nervous system imbalance have also been suggested.

Lung Disease

Research has demonstrated increased risk of developing asthma and COPD from increased exposure to traffic-related air pollution. Additionally, air pollution has been associated with increased hospitalization and mortality from asthma and COPD. Chronic obstructive pulmonary disease (COPD) includes diseases such as chronic bronchitis and emphysema.

A study conducted in 1960–1961 in the wake of the Great Smog of 1952 compared 293 London residents with 477 residents of Gloucester, Peterborough, and Norwich, three towns with low reported death rates from chronic bronchitis. All subjects were male postal truck drivers aged 40 to 59. Compared to the subjects from the outlying towns, the London subjects exhibited more severe respiratory symptoms (including cough, phlegm, and dyspnea), reduced lung function (FEV_1 and peak flow rate), and increased sputum production and purulence. The differences were more pronounced for subjects aged 50 to 59. The study controlled for age and smoking habits, so concluded that air pollution was the most likely cause of the observed differences. More recent studies have shown that air pollution exposure from traffic reduces lung function development in children and lung function may be compromised by air pollution even at low concentrations. Air pollution exposure also cause lung cancer in non smokers.

It is believed that much like cystic fibrosis, by living in a more urban environment serious health hazards become more apparent. Studies have shown that in urban areas patients suffer mucus hypersecretion, lower levels of lung function, and more self-diagnosis of chronic bronchitis and emphysema.

Cancer (Lung Cancer)

Unprotected exposure to PM2.5 air pollution can be equivalent to smoking multiple cigarettes per day, potentially increasing the risk of Cancer, which is mainly the result of environmental factors.

A review of evidence regarding whether ambient air pollution exposure is a risk factor for cancer in 2007 found solid data to conclude that long-term exposure to PM2.5 (fine particulates) increases the overall risk of non-accidental mortality by 6% per a 10 microg/m^3 increase. Exposure to PM2.5 was also associated with an increased risk of mortality from lung cancer (range: 15% to 21% per 10 microg/m^3 increase) and total cardiovascular mortality (range: 12% to 14% per a 10 microg/m^3 increase). The review further noted that living close to busy traffic appears to be associated with elevated risks of these three outcomes – increase in lung cancer deaths, cardiovascular deaths, and overall non-accidental deaths. The reviewers also found suggestive evidence that exposure to PM2.5 is positively associated with mortality from coronary heart diseases and exposure to SO_2 increases mortality from lung cancer, but the data was insufficient to provide solid conclusions. Another investigation showed that higher activity level increases deposition fraction of aerosol particles in human lung and recommended avoiding heavy activities like running in outdoor space at polluted areas.

In 2011, a large Danish epidemiological study found an increased risk of lung cancer for patients who lived in areas with high nitrogen oxide concentrations. In this study, the association was higher for non-smokers than smokers. An additional Danish study, also in 2011, likewise noted evidence of possible associations between air pollution and other forms of cancer, including cervical cancer and brain cancer.

In December 2015, medical scientists reported that cancer is overwhelmingly a result of environmental factors, and not largely down to bad luck. Maintaining a healthy weight, eating a healthy diet, minimizing alcohol and eliminating smoking reduces the risk of developing the disease, according to the researchers.

Children

In the United States, despite the passage of the Clean Air Act in 1970, in 2002 at least 146 million Americans were living in non-attainment areas—regions in which the concentration of certain air pollutants exceeded federal standards. These dangerous pollutants are known as the criteria pollutants, and include ozone, particulate matter, sulfur dioxide, nitrogen dioxide, carbon monoxide, and lead. Protective measures to ensure children's health are being taken in cities such as New Delhi, India where buses now use compressed natural gas to help eliminate the "pea-soup" smog. A recent study in Europe has found that exposure to ultrafine particles can increase blood pressure in children. According to a WHO report-2018, polluted air is a main cause poisoning millions of children under the age of 15 years and ruining their lives which resulting to death of some six hundred thousand children annually.

Infants

Ambient levels of air pollution have been associated with preterm birth and low birth weight. A 2014 WHO worldwide survey on maternal and perinatal health found a statistically significant association between low birth weights (LBW) and increased levels of exposure to PM2.5. Women in regions with greater than average PM2.5 levels had statistically significant higher odds of pregnancy resulting in a low-birth weight infant even when adjusted for country-related variables. The effect is thought to be from stimulating inflammation and increasing oxidative stress.

A study by the University of York found that in 2010 exposure to PM2.5 was strongly associated with 18% of preterm births globally, which was approximately 2.7 million premature births. The countries with the highest air pollution associated preterm births were in South and East Asia, the Middle East, North Africa, and West sub-Saharan Africa.

The source of PM 2.5 differs greatly by region. In South and East Asia, pregnant women are frequently exposed to indoor air pollution because of the wood and other biomass fuels used for cooking which are responsible for more than 80% of regional pollution. In the Middle East, North Africa and West sub-Saharan Africa, fine PM comes from natural sources, such as dust storms. The United States had an estimated 50,000 preterm births associated with exposure to PM2.5 in 2010.

A study performed by Wang, et al. between the years of 1988 and 1991 has found a correlation between sulfur Dioxide (SO_2) and total suspended particulates (TSP) and preterm births and low birth weights in Beijing. A group of 74,671 pregnant women, in four separate regions of Beijing, were monitored from early pregnancy to delivery along with daily air pollution levels of sulfur Dioxide and TSP (along with other particulates). The estimated reduction in birth weight was 7.3 g for every 100 $\mu g/m^3$ increase in SO_2 and 6.9g for each 100 $\mu g/m^3$ increase in TSP. These associations were statistically significant in both summer and winter, although, summer was greater. The proportion of low birth weight attributable to air pollution, was 13%. This is the largest attributable risk ever reported for the known risk factors of low birth weight. Coal stoves, which are in 97% of homes, are a major source of air pollution in this area.

Brauer et al. studied the relationship between air pollution and proximity to a highway with pregnancy outcomes in a Vancouver cohort of pregnant woman using addresses to estimate exposure during pregnancy. Exposure to NO, NO_2, CO PM10 and PM2.5 were associated with infants born small for gestational age (SGA). Women living <50meters away from an expressway or highway were 26% more likely to give birth to a SGA infant.

Clean Areas

Even in the areas with relatively low levels of air pollution, public health effects can be significant and costly, since a large number of people breathe in such pollutants. A study published in 2017 found that even in areas of the U.S. where ozone and PM2.5 meet federal standards, Medicare recipients who are exposed to more air pollution have higher mortality rates. A 2005 scientific study for the British Columbia Lung Association showed that a small improvement in air quality (1% reduction of ambient PM2.5 and ozone concentrations) would produce $29 million in annual savings in the Metro Vancouver region in 2010. This finding is based on health valuation of lethal (death) and sub-lethal (illness) affects.

Central Nervous System

Data is accumulating that air pollution exposure also affects the central nervous system. It was discovered that early exposure to air pollution causes the same damaging changes in the brain as autism and schizophrenia. The study also shows that air pollution also affected short-term memory, learning ability, and impulsivity. Lead researcher Professor Deborah Cory-Slechta said that "When we looked closely at the ventricles, we could see that the white matter that normally surrounds them hadn't fully developed. It appears that inflammation had damaged those brain cells and prevented that region of the brain from developing, and the ventricles simply expanded to fill the space. Our findings add to the growing body of evidence that air pollution may play a role in autism, as well as in other neurodevelopmental disorders." In a study of mice, air pollution also has a more significant negative effect on males than on females.

In 2015, experimental studies reported the detection of significant episodic (situational) cognitive impairment from impurities in indoor air breathed by test subjects who were not informed about changes in the air quality. Researchers at the Harvard University and SUNY Upstate Medical University and Syracuse University measured the cognitive performance of 24 participants in three different controlled laboratory atmospheres that simulated those found in "conventional" and "green" buildings, as well as green buildings with enhanced ventilation. Performance was evaluated objectively using the widely used Strategic Management Simulation software simulation tool, which is a well-validated assessment test for executive decision-making in an unconstrained situation allowing initiative and improvisation. Significant deficits were observed in the performance scores achieved in increasing concentrations of either volatile organic compounds (VOCs) or carbon dioxide, while keeping other factors constant. The highest impurity levels reached are not uncommon in some classroom or office environments. Air pollution increases the risk of dementia in people over 50 years old.

Agricultural Effects

In India in 2014, it was reported that air pollution by black carbon and ground level ozone had reduced crop yields in the most affected areas by almost half in 2011 when compared to 1980 levels.

Alternatives to Pollution

There are now practical alternatives to the principal causes of air pollution:

- Areas downwind (over 20 miles) of major airports more than double *total particulate emissions in air*, even when factoring in areas with frequent ship calls, and heavy freeway and city traffic like Los Angeles. Aviation biofuel mixed in with jetfuel at a 50/50 ratio can reduce jet derived cruise altitude particulate emissions by 50–70%, according to a NASA led 2017 study (however, this should imply ground level benefits to urban air pollution as well).
- Ship propulsion and idling can be switched to much cleaner fuels like natural gas. (Ideally a renewable source but not practical yet.)
- Combustion of fossil fuels for space heating can be replaced by using ground source heat pumps and seasonal thermal energy storage.
- Electric power generation from burning fossil fuels can be replaced by power generation from nuclear and renewables. For poor nations, heating and home stoves that contribute much to regional air pollution can be replaced by a much cleaner fossil fuel like natural gas, or ideally, renewables.

- Motor vehicles driven by fossil fuels, a key factor in urban air pollution, can be replaced by electric vehicles. Though lithium supply and cost is a limitation, there are alternatives. Herding more people into clean public transit such as electric trains can also help. Nevertheless, even in emission-free electric vehicles, rubber tires produce significant amounts of air pollution themselves, ranking as 13th worst pollutant in Los Angeles.
- Reducing travel in vehicles can curb pollution. After Stockholm reduced vehicle traffic in the central city with a congestion tax, nitrogen dioxide and PM10 pollution declined, as did acute pediatric asthma attacks.
- Biodigesters can be utilized in poor nations where slash and burn is prevalent, turning a useless commodity into a source of income. The plants can be gathered and sold to a central authority that will break it down in a large modern biodigester, producing much needed energy to use.
- Induced humidity and ventilation both can greatly dampen air pollution in enclosed spaces, which was found to be relatively high inside subway lines due to braking and friction and relatively less ironically inside transit buses than lower sitting passenger automobiles or subways.

Reduction Efforts

Various pollution control technologies and strategies are available to reduce air pollution. At its most basic level, land-use planning is likely to involve zoning and transport infrastructure planning. In most developed countries, land-use planning is an important part of social policy, ensuring that land is used efficiently for the benefit of the wider economy and population, as well as to protect the environment.

Because a large share of air pollution is caused by combustion of fossil fuels such as coal and oil, the reduction of these fuels can reduce air pollution drastically. Most effective is the switch to clean power sources such as wind power, solar power, hydro power which don't cause air pollution. Efforts to reduce pollution from mobile sources includes primary regulation (many developing countries have permissive regulations),expanding regulation to new sources (such as cruise and transport ships, farm equipment, and small gas-powered equipment such as string trimmers, chainsaws, and snowmobiles), increased fuel efficiency (such as through the use of hybrid vehicles), conversion to cleaner fuels or conversion to electric vehicles.

Titanium dioxide has been researched for its ability to reduce air pollution. Ultraviolet light will release free electrons from material, thereby creating free radicals, which break up VOCs and NOx gases. One form is superhydrophilic.

In 2014, Prof. Tony Ryan and Prof. Simon Armitage of University of Sheffield prepared a 10 meter by 20 meter-sized poster coated with microscopic, pollution-eating nanoparticles of titanium dioxide. Placed on a building, this giant poster can absorb the toxic emission from around 20 cars each day.

A very effective means to reduce air pollution is the transition to renewable energy. According to a study published in Energy and Environmental Science in 2015 the switch to 100% renewable energy in the United States would eliminate about 62,000 premature mortalities per year and about 42,000 in 2050, if no biomass were used. This would save about $600 billion in health costs a year due to reduced air pollution in 2050, or about 3.6% of the 2014 U.S. gross domestic product.

Control Devices

The following items are commonly used as pollution control devices in industry and transportation. They can either destroy contaminants or remove them from an exhaust stream before it is emitted into the atmosphere.

- Particulate control:
    - Mechanical collectors (dust cyclones, multicyclones).
    - Electrostatic precipitators: An electrostatic precipitator (ESP), or electrostatic air cleaner is a particulate collection device that removes particles from a flowing gas (such as air), using the force of an induced electrostatic charge. Electrostatic precipitators are highly efficient filtration devices that minimally impede the flow of gases through the device, and can easily remove fine particulates such as dust and smoke from the air stream.
    - Baghouses: Designed to handle heavy dust loads, a dust collector consists of a blower, dust filter, a filter-cleaning system, and a dust receptacle or dust removal system (distinguished from air cleaners which utilize disposable filters to remove the dust).
    - Particulate scrubbers: Wet scrubber is a form of pollution control technology. The term describes a variety of devices that use pollutants from a furnace flue gas or from other gas streams. In a wet scrubber, the polluted gas stream is brought into contact with the scrubbing liquid, by spraying it with the liquid, by forcing it through a pool of liquid, or by some other contact method, so as to remove the pollutants.
- Scrubbers:
    - Baffle spray scrubber.
    - Cyclonic spray scrubber.
    - Ejector venturi scrubber.
    - Mechanically aided scrubber.

- Spray tower.
- Wet scrubber.

- NO_X control:
  - Low NO_X burners.
  - Selective catalytic reduction (SCR).
  - Selective non-catalytic reduction (SNCR).
  - NOx scrubbers.
  - Exhaust gas recirculation.
  - Catalytic converter (also for VOC control).
- VOC abatement:
  - Adsorption systems, using activated carbon, such as Fluidized Bed Concentrator.
  - Flares.
  - Thermal oxidizers.
  - Catalytic converters.
  - Biofilters.
  - Absorption (scrubbing).
  - Cryogenic condensers.
  - Vapor recovery systems.
- Acid Gas/SO_2 control:
  - Wet scrubbers.
  - Dry scrubbers.
  - Flue-gas desulfurization.
- Mercury control:
  - Sorbent Injection Technology.
  - Electro-Catalytic Oxidation (ECO).
  - K-Fuel.

- Dioxin and furan control.
- Miscellaneous associated equipment:
  - Source capturing systems.
  - Continuous emissions monitoring systems (CEMS).

Hotspots

Air pollution hotspots are areas where air pollution emissions expose individuals to increased negative health effects. They are particularly common in highly populated, urban areas, where there may be a combination of stationary sources (e.g. industrial facilities) and mobile sources (e.g. cars and trucks) of pollution. Emissions from these sources can cause respiratory disease, childhood asthma, cancer, and other health problems. Fine particulate matter such as diesel soot, which contributes to more than 3.2 million premature deaths around the world each year, is a significant problem. It is very small and can lodge itself within the lungs and enter the bloodstream. Diesel soot is concentrated in densely populated areas, and one in six people in the U.S. live near a diesel pollution hot spot.

While air pollution hotspots affect a variety of populations, some groups are more likely to be located in hotspots. Previous studies have shown disparities in exposure to pollution by race and income. Hazardous land uses (toxic storage and disposal facilities, manufacturing facilities, major roadways) tend to be located where property values and income levels are low. Low socioeconomic status can be a proxy for other kinds of social vulnerability, including race, a lack of ability to influence regulation and a lack of ability to move to neighborhoods with less environmental pollution. These communities bear a disproportionate burden of environmental pollution and are more likely to face health risks such as cancer or asthma.

Studies show that patterns in race and income disparities not only indicate a higher exposure to pollution but also higher risk of adverse health outcomes. Communities characterized by low socioeconomic status and racial minorities can be more vulnerable to cumulative adverse health impacts resulting from elevated exposure to pollutants than more privileged communities. Blacks and Latinos generally face more pollution than whites and Asians, and low-income communities bear a higher burden of risk than affluent ones. Racial discrepancies are particularly distinct in suburban areas of the Southern United States and metropolitan areas of the Midwestern and Western United States. Residents in public housing, who are generally low-income and cannot move to healthier neighborhoods, are highly affected by nearby refineries and chemical plants.

Cities

Air pollution is usually concentrated in densely populated metropolitan areas, especially in developing countries where environmental regulations are relatively lax or nonexistent.

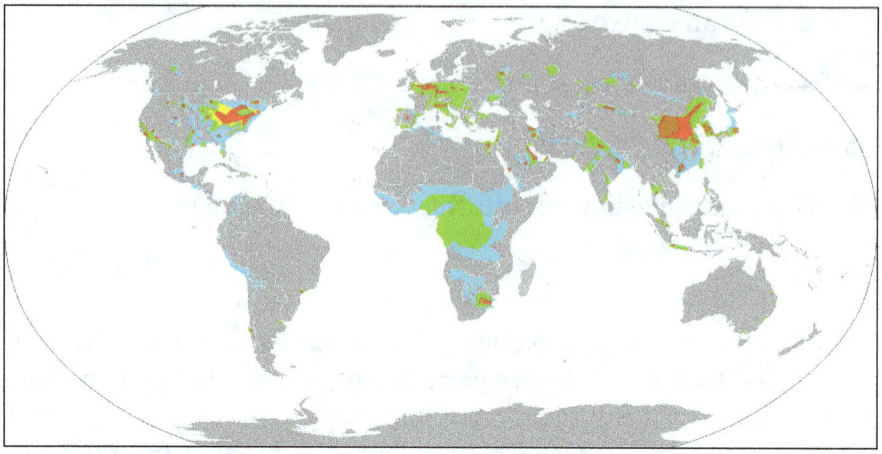

Nitrogen dioxide concentrations as measured from satellite 2002–2004.

However, even populated areas in developed countries attain unhealthy levels of pollution, with Los Angeles and Rome being two examples. Between 2002 and 2011 the incidence of lung cancer in Beijing near doubled. While smoking remains the leading cause of lung cancer in China, the number of smokers is falling while lung cancer rates are rising.

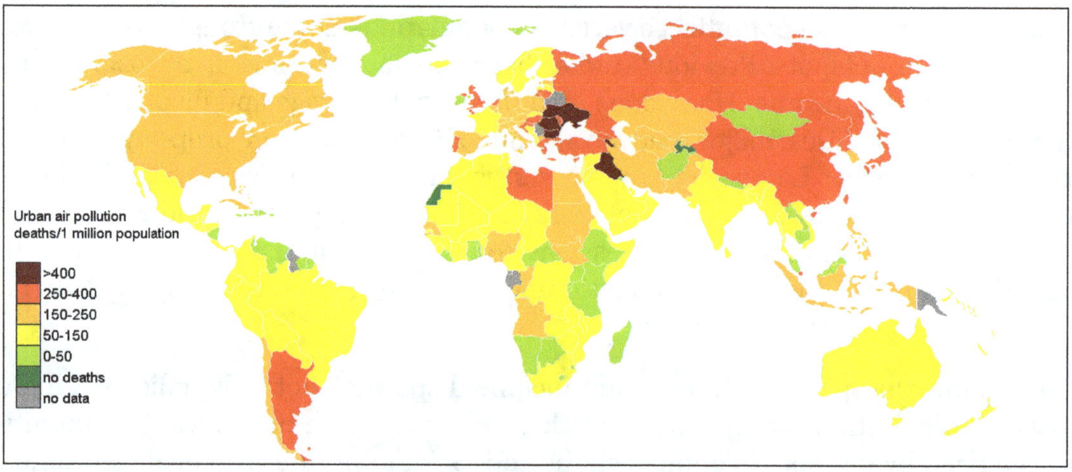

Deaths from air pollution in 2004.

Most polluted cities by PM	
Particulate matter, μg/m³ (2004)	City
168	Cairo, Egypt
150	Delhi, India
128	Kolkata, India (Calcutta)
125	Tianjin, China
123	Chongqing, China
109	Kanpur, India
109	Lucknow, India
104	Jakarta, Indonesia
101	Shenyang, China

Water Chemistry

Water is a unique substance in our discovered universe. The presence of water on earth is in itself unique, for the planet earth has few natural liquids. Water is the prime resource of man's food supply and his most important household and industrial tool. But most important is the fact that water is a major constituent of all living matter, comprising up to two-thirds of the human body. Next to the air we breathe, water is mankind's most important substance.

Physical and Chemical Changes

Physical change: A physical change is one in which there is no change in the molecules which make up a given substance. Turning water into ice or vapor does not constitute a chemical change because the same molecules make up the liquid, solid and vapor states of water. The only difference between ice, steam and water is this: Molecules in ice essentially have no freedom. The only vibrate within the crystal. The molecules in water are free to move within the limits of the container, as limited by gravity. The molecules in stream are completely free to move within the container, if any. They are essentially unaffected by gravity.

Chemical change: A chemical change occurs when new molecules are formed as a result of the change.

When water turns to steam at 212 °F, a physical change occurs. On the other hand, when propane gas is ignited, it turns to carbon dioxide and water vapor in a chemical change. And change, of course, continues all the time. Let's briefly consider the types of change by examine the compounds and mixtures.

Compounds and Mixtures

How can one distinguish between compounds and mixtures? A compound has a definite and unvarying composition.

Water is a typical compound. It is composed of two elements hydrogen and oxygen in definite proportions. Regardless of where one ands water, it always consists of these two elements and always in the same proportion. Salt is another common compound.

Whether it comes from a salt mine or is produced in a laboratory, salt is a compound of the two elements sodium and chlorine in an unvarying ratio.

Water, as a typical compound, also suggests another characteristic of the compound, namely a unique "personality" of its own. Although made up of hydrogen and oxygen, water is quite different from these two elements both physically and chemically. And so we should add to our definitions: A compound has well defined characteristics of its own, usually entirely different from those of its component elements. Further, water freezes at 32 °F and boils at 212 °F. This indicates another characteristic of the compound: a pure compound has a definite freezing and a definite boiling point.

And finally, water, as a typical compound, is a uniform substance no matter whether one is considering a drop, a glassful or a lake of it. Thus, a compound is homogeneous.

In sharp contrast, a mixture will vary in the amount of the ingredients it contains. A mixture of sand and salt, for example, may have a bit of salt and a large amount of sand. Or it may be a blend of a large amount of salt and sand. No exact ratios of substances are necessary to constitute a mixture. At the same the ingredients in a mixture continue to maintain their essential properties. The salt still tastes salty; the sand continues to be gritty. The properties of the mixture are simply the total of the separate properties of the salt and sand. In this salt sand mixture the original ingredients could be recovered through some type of mechanical process. And finally, a mixture may have varying proportions of its ingredients in different parts of the sample. There may be more salt than sand at the bottom and less at the top of a mixture. In a word, mixtures are usually heterogeneous.

Over the years scientists have discovered in nature or produced in their laboratories more than 100 elements. Each of these basic materials has its own peculiar atomic structure. The molecule of an element may consist of one atom or two or more similar atoms. These atoms are inconceivably small. The weight of an oxygen atom, for example, has been determined at 0.000,000,000,000,000,000,000,0266 gram. (There are 453.6 grams in a pound.) Obviously, ordinary units of measurement are insufficient for use in determining the weight of an atom or a molecule.

Atom

The smallest possible unit of an element. The atom can only be broken down into the fundamental particles of matter, such as the proton, electron, neutron, antiproton and meson.

Molecules

Molecules consist of two or more atoms of one or more elements. (Cl_2) is a molecule of chlorine. ($NaCl$) is a molecule of salt.

Gram

The gram is the basic unit of weight in the metric system. It is equal to 1/28 of an ounce (.0022046 pound or 15.4324 grains troy). It is meant to be, and virtually is, the weight of distilled water at 4 °C contained in a cube whose edge is one hundredth of a meter (one cubic centimeter). 453.6 g = 1 pound.

As it occurs in nature, oxygen is a mixture of three isotopes-oxygen-16, oxygen-17, and oxygen-18. Over the years, chemists developed a table of atomic weights, with a value of 16 established for this natural mixture. However, because of different needs, physicists developed a separate table based upon the single isotope, oxygen-16. Thus, minor but definite variations existed between the tables. Recently a new table based upon carbon-12 has been adopted, which apparently will be used by both. The atomic weights used in this course are taken from the new table.

Over a century ago, chemists developed a purely relative scale for atomic weights. At the time, they assigned a mass of 16 to the oxygen atom. They then expressed the atomic weight for all other elements in terms relative to the weight of 16 assigned to the oxygen atom.

The hydrogen atom was discovered to weigh approximately 1/16 that of oxygen. Its actual atomic weight is 1.00797.

The weight of molecules can be easily calculated if the formula for a given molecule is known. Simply add the atomic weights of the various atoms that make up the molecule. Here are a few examples of how molecular weight is determined.

Substance/Formula	Computation of Molecular Weight
Oxygen molecule/O_2	2 x atomic weight = 2 x 16 = 32
Hydrogen molecule/H_2	2 x 1 = 2
Water molecule/H_2O	(2 x 1) + 16 = 18

Biochemical Properties of Water

Water is tasteless, odorless, and transparent. In small quantities, it is also colorless. However, when a large amount of water is observed, as in a lake or the ocean, it is actually light blue in color. The blue hue of water is an intrinsic property and is caused by selective absorption and scattering of white light. These and other properties of water depend on its chemical structure. The transparency of water is important for organisms that live in water. Because water is transparent, sunlight can pass through it. Sunlight is needed by water plants and other water organisms for photosynthesis.

Chemical Structure of Water

Each molecule of water consists of one atom of oxygen and two atoms of hydrogen, so it has the chemical formula H_2O. The arrangement of atoms in a water molecule, shown in figure below, explains many of water's chemical properties. In each water molecule, the nucleus of the oxygen atom (with 8 positively charged protons) attracts electrons much more strongly than do the hydrogen nuclei (with only one positively charged proton). This results in a negative electrical charge near the oxygen atom (due to the "pull" of the negatively charged electrons toward the oxygen nucleus) and a positive electrical charge near the hydrogen atoms. A difference in electrical charge between different parts of a molecule is called polarity. A polar molecule is a molecule in which part of the molecule is positively charged and part of the molecule is negatively charged.

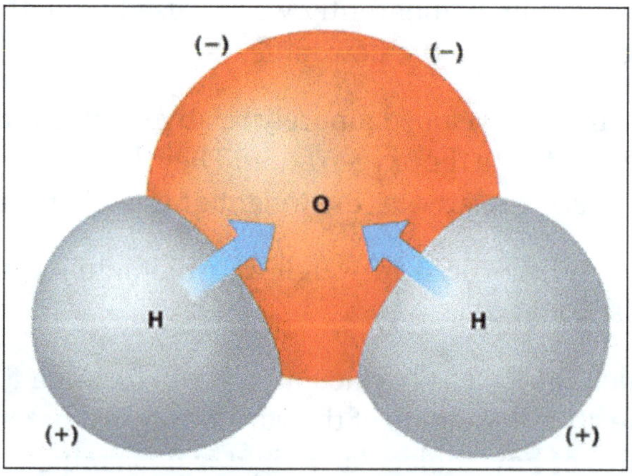

This model shows the arrangement of oxygen and hydrogen atoms in a water molecule. A water molecule has a bent or angular (non-linear) shape, with an angle of about 105°. The nucleus of the oxygen atom attracts electrons more strongly than do the hydrogen nuclei. As a result, the middle part of the molecule near oxygen has a negative charge, and the other parts of the molecule have a positive charge. In essence, the electrons are "pulled" toward the nucleus of the oxygen atom and away from the hydrogen atom nuclei. Water is a polar molecule, with an unequal distribution of charge throughout the molecule.

This model is an atomic diagram of water, showing the two hydrogen atoms and oxygen atom in the center. The protons (red) are in the center (nucleus) of each atom, and the electrons (light blue) circle each nucleus.

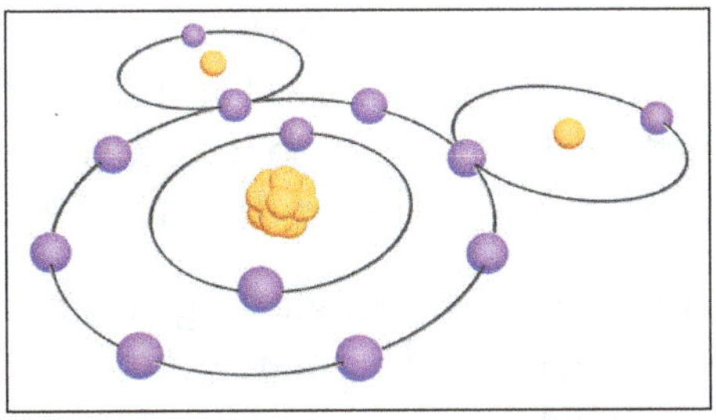

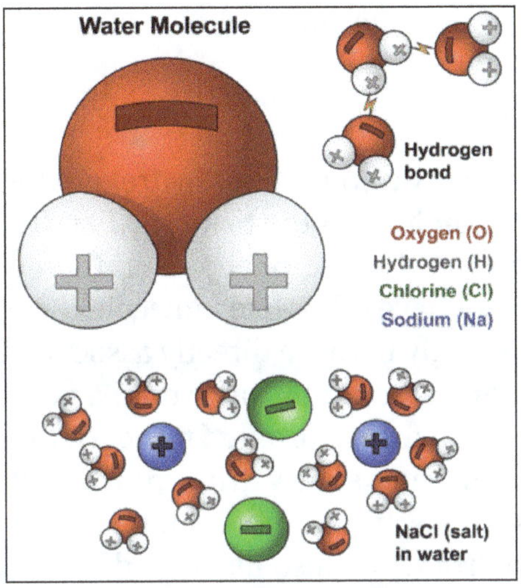

This diagram shows the positive and negative parts of a water molecule. It also depicts how a charge, such as on an ion (Na or Cl, for example) can interact with a water molecule.

Hydrogen Bonding

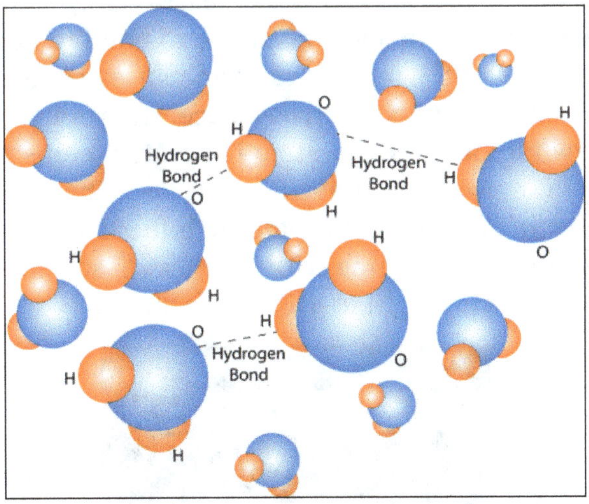

Hydrogen bonds form between positively and negatively charged parts of water molecules. The bonds hold the water molecules together. How do you think this might affect water's properties?

Opposite electrical charges attract one another. Therefore, the positive part of one water molecule is attracted to the negative parts of other water molecules. Because of this attraction, bonds form between hydrogen and oxygen atoms of adjacent water molecules, as demonstrated in figure below. This type of bond always involves a hydrogen atom, so it is called a hydrogen bond. Hydrogen bonds are bonds between molecules, and they are not as strong as bonds within molecules. Nonetheless, they help hold water molecules together.

Hydrogen bonds can also form within a single large organic molecule. For example, hydrogen bonds that form between different parts of a proteinmolecule bend the molecule into a distinctive shape, which is important for the protein's functions. Hydrogen bonds also hold together the two nucleotide chains of a DNA molecule.

Sticky and Wet Water

Water has some unusual properties due to its hydrogen bonds. One property is cohesion, the tendency for water molecules to stick together. The cohesive forces between water molecules are responsible for the phenomenon known as surface tension. The molecules at the surface do not have other like molecules on all sides of them and consequently they cohere more strongly to those directly associated with them on the surface. For example, if you drop a tiny amount of water onto a very smooth surface, the water molecules will stick together and form a droplet, rather than spread out over the surface. The same thing happens when water slowly drips from a leaky faucet. The water doesn't fall from the faucet as individual water molecules but as droplets of water. The tendency of water to stick together in droplets is also illustrated by the dew drops in figure.

Another important physical property of water, is adhesion. In terms of water, adhesion is the bonding of a water molecule to another substance, such as the sides of a leaf's veins. This process happens because hydrogen bonds are special in that they break and reform with great frequency. This constant rearranging of hydrogen bonds allows a percentage of all the molecules in a given sample to bond to another substance. This grip-like characteristic that water molecules form causes capillary action, the ability of a liquid to flow against gravity in a narrow space. An example of capillary action is when you place a straw into a glass of water. The water seems to climb up the straw before you even place your mouth on the straw. The water has created hydrogen bonds with the surface of the straw, causing the water to adhere to the sides of the straw. As the hydrogen bonds keep interchanging with the straw's surface, the water molecules interchange positions and some begin to ascend the straw.

Droplets of dew cling to a spider web, demonstrating cohesion, the tendency of water molecules to stick together because of hydrogen bonds.

Adhesion and capillary action are necessary to the survival of most organisms. It is the mechanism that is responsible for water transport in plants through roots and stems, and in animals through small blood vessels.

Hydrogen bonds also explain why water's boiling point (100 °C) is higher than the boiling points of similar substances without hydrogen bonds. Because of water's relatively high boiling point, most water exists in a liquid state on Earth. Liquid water is needed by all living organisms. Therefore, the availability of liquid water enables life to survive over much of the planet.

Furthermore, water has a high specific heat because it takes a lot of energy to raise or lower the temperature of water. As a result, water plays a very important role in temperature regulation. Since cells are made up of water, this property helps to maintain homeostasis.

Density of Ice and Water

The melting point of water is 0 °C. Below this temperature, water is a solid(ice). Unlike most chemical substances, water in a solid state has a lower density than water in a liquid state. This is because water expands when it freezes. Again, hydrogen bonding is the reason. Hydrogen bonds cause water molecules to line up less efficiently in ice than in liquid water. As a result, water molecules are spaced farther apart in ice, giving ice a lower density than liquid water. A substance with lower density floats on a substance with higher density. This explains why ice floats on liquid water, whereas many other solids sink to the bottom of liquid water.

In a large body of water, such as a lake or the ocean, the water with the greatest density always sinks to the bottom. Water is most dense at about 4 °C. As a result, the water at the bottom of a lake or the ocean usually has temperature of about 4 °C. In climates with cold winters, this layer of 4 °C water insulates the bottom of a lake from freezing temperatures. Lake organisms such as fish can survive the winter by staying in this cold, but unfrozen, water at the bottom of the lake.

Analysis of Water Chemistry

Water chemistry analyses are carried out to identify and quantify the chemical components and properties of water samples. The type and sensitivity of the analysis depends on the purpose of the analysis and the anticipated use of the water. Chemical water analysis is carried out on water used in industrial processes, on waste-water stream, on rivers and stream, on rainfall and on the sea. In all cases the results of the analysis provides information that can be used to make decisions or to provide re-assurance that conditions are as expected. The analytical parameters selected are chosen to be appropriate for the decision making process or to establish acceptable normality. Water chemistry analysis is often the groundwork of studies of water quality, pollution, hydrology and geothermal waters. Analytical methods routinely used can detect and measure all the natural elements and their inorganic compounds and a very wide range of organic chemical species using methods such as gas chromatography and mass spectrometry. In water treatment plants producing drinking water and in some industrial processes using products with distinctive taste and odours, specialised organoleptic methods may be used to detect smells at very low concentrations.

Environmental Water

An EPA scientist samples water in Florida Everglades.

Samples of water from the natural environment are routinely taken and analysed as part of a pre-determined monitoring programme by regulatory authorities to ensure that waters remain unpolluted, or if polluted, that the levels of pollution are not increasing or are falling in line with an agreed remediation plan. An example of such a scheme is the harmonised monitoring scheme operated on all the major river systems in the UK. The parameters analysed will be highly dependent on nature of the local environment and the polluting sources in the area. In many cases the parameters will reflect the national and local water quality standards determined by law or other regulations. Typical parameters for ensuring that unpolluted surface waters remain within acceptable chemical standards include pH, major cations and anions including ammonia, nitrate, nitrite, phosphate, conductivity, phenol, chemical oxygen demand (COD) and biochemical oxygen demand (BOD).

Drinking Water Supplies

Surface or ground water abstracted for the supply of drinking water must be capable of meeting rigorous chemical standards following treatment. This requires a detailed knowledge of the water entering the treatment plant. In addition to the normal suite of environmental chemical parameters, other parameters such as hardness, phenol, oil and in some cases a real-time organic profile of the incoming water as in the River Dee regulation scheme.

Industrial Process Water

In industrial process, the control of the quality of process water can be critical to the quality of the end product. Water is often used as a carrier of reagents and the loss of reagent to product must be continuously monitored to ensure that correct replacement rate. Parameters measured relate specifically to the process in use and to any of the expected contaminants that may arise as by-products. This may include unwanted organic chemicals appearing in an inorganic chemical process through contamination with oils and greases from machinery. Monitoring the quality of the waste water discharged from industrial premises is a key factor in controlling and minimising pollution of the environment. In this application monitoring schemes analyse for all possible contaminants arising within the process and in addition contaminants that may have particularly adverse impacts on the environment such as cyanide and many organic species such as pesticides. In then nuclear industry analysis focuses on specific isotopes or elements of interest where the nuclear industry makes waste water discharges to rivers which have drinking water abstraction on them, radio-isotopes which could potentially be harmful or those with long half-lives such as tritium will form part of the routine monitoring suite.

Forensic Analysis

In environmental management, water analysis is frequently deployed when pollution is suspected to identify the pollutant in order to take remedial action. The analysis can often enable the polluter to be identified. Such forensic work can examine the ratios of various components and can "type" samples of oils or other mixed organic contaminants to directly link the pollutant with the source. In drinking water supplies the cause of unacceptable quality can similarly be determined by carefully targeted chemical analysis of samples taken throughout the distribution system. In manufacturing, off-spec products may be directly tied back to unexpected changes in wet processing stages and analytical chemistry can identify which stages may be at fault and for what reason.

Methodology

To ensure consistency and repeatability, the methods use in the chemical analysis of water samples are often agreed and published at a national or state level. By convention these are often referred to as "Blue book"s.

The methods defined in the relevant standards can be broadly classified as:

- *Conventional wet chemistry* including the Winkler method for dissolved oxygen, precipitation, filtration for solids, acidification, neutralisation titration etc. Colourimetric methods such as MBAS assay which indicates anionic surfactants in water and on site comparator methods to determine chlorine and chloramines. Nephelometers are used to measure solids concentrations as turbidity. These methods are generally robust and well tried and inexpensive, giving a reasonable degree of accuracy at modest sensitivity.

- *Electro chemistry* including pH, conductivity and dissolved oxygen using oxygen electrode. These methods yield accurate and precise results using electronic equipment capable of feeding results directly into a laboratory data management system.
- Spectrophotometry is used particularly for metallic elements in solution producing results with very high sensitivity, but which may require some sample preparation prior to analysis and may also need specialised sampling methods to avoid sample deterioration in transit.
- Chromatography is used for many organic species which are volatile or which can yield a characteristic volatile component of after initial chemical processing.
- Ion chromatography is a sensitive and stable technique that can measure lithium, ammonium NH_4 and many other low molecular weight ions using ion exchange technology.
- Gas chromatography can be used to determine methane, carbon dioxide, cyanide, oxygen, nitrogen and many other volatile components at reasonable sensitivities.
- Mass spectrometry is used where very high sensitivity is required and is sometimes used as a back-end process after gas liquid chromatography for detecting trace organic chemicals.

Depending on the components, different methods are applied to determine the quantities or ratios of the components. While some methods can be performed with standard laboratory equipment, others require advanced devices, such as inductively coupled plasma mass spectrometry (ICP-MS).

Water Pollution

Water pollution occurs when undesirable foreign substances are introduced into natural water. The substances may be chemical or biological in nature. Common pollutants include human or animal waste; disease-producing organisms; radioactive materials; toxic metals such as lead or mercury; agricultural chemicals such as pesticides, herbicides, or fertilizers; acid rain; and high-temperature water discharged from power plants, often called "thermal pollution." Pollutants in water are dangerous for human or animal consumption and harm crops. High temperatures may cause algae to grow rapidly, rendering water unfit for consumption.

Point sources of pollution, such as an oil leak from a pipeline or chemical waste from a factory, can often be controlled. Nonpoint sources, such as runoff sediment and nitrate-rich water from feedlots represent larger amounts of pollution and are difficult to identify and remedy. Pollution from nonpoint sources may pass into streams or aquifers, covering a wide area.

Although water has been identified on several planets, none has as much water as Earth, of which 70 percent is covered with water. Approximately 97.4 percent of the water on Earth is found in oceans and is too salty for human consumption. An additional 2.6 percent is freshwater found in underground bodies of water called aquifers or frozen in glaciers or polar ice caps. Less than 0.02 percent of Earth's water is present in lakes, rivers, or the atmosphere.

In a few places, water is pure enough to drink directly from wells or springs, but increasingly water must be treated to remove dangerous contaminants, and substances such as chlorine, chloramines, or ozone must be added to kill harmful bacteria.

Pollutants in water are commonly measured and reported as parts per million (ppm) or parts per billion (ppb). A solution that contains 2 grams(0.071 ounces) of lead in 1 million grams (2,205 pounds) of water (1,000 liters, or 264.2 gallons) is a 2 ppm solution. A 1 ppb solution of calcium contains 1 gram (0.036 ounces) of calcium in 1 billion grams (2,205,000 pounds) of water. A concentration of 1 ppm is the same as 1 milligram(3.6×10^{-5} ounces) per liter.

While it is impractical to remove all impurities from water, the Safe Drinking Water Act, passed by the U.S. Congress in 1974, gives the Environmental Protection Agency (EPA) the authority to set limits for harmful contaminants in water. For each substance, the EPA establishes Maximum Contaminant Level Goals (MCLGs), levels at which the substance can be consumed over a long period of time with no known adverse effects. This level is defined as the amount of impurity that could be present in two liters of water drunk by a person weighing 70 kilograms (154 pounds), each day for seventy years, without ill effects. In addition, the EPA sets Maximum Contaminant Levels (MCLs) of substances for exposure at any single time. A single exposure to concentrations of pollutants below the MCL is considered to be harmless. The MCLG of lead is 0; continuous exposure to lead in any concentration is considered hazardous. The MCL of lead is 0.015 ppm. Both the MCLG and MCL of mercury are set at 0.002 ppm.

Specialized analytical equipment allows technicians to monitor pollutants. In the field, pH meters are used to measure acidity and turbidometers measure the presence of suspended solids. Samples taken to laboratories are analyzed by gas chromatography to determine the presence of organic. Compounds such as vinyl chloride, by emission spectroscopy to detect heavy metals, and by high performance liquid chromatography (HPLC) to detect pesticide residues. Such instruments are capable of detecting as little as one part per trillion of pollutants in water.

For much of history, humans used waterways and bodies of water as waste dumps. When the human population was low, fewer people were exposed to the effects of pollution, and the sources were fewer and produced less pollution. During the Industrial Revolution of the nineteenth century, water pollution was recognized as a danger to public health.

Even early settlers were concerned with water quality. Two hundred years before laws were written to protect consumers from lead poisoning, Benjamin Franklin wrote of a family that suffered gastrointestinal pains after drinking water collected from their lead roof. During the trek west, members of wagon trains avoided drinking from stagnant pools, some of which contained large amounts of alkali.

As populations and production grew, industrial and household refuse accumulated, and it became clear that many discarded materials did not simply disappear, but were spread through the water table, absorbed by lower forms of life and passed up the food chain, causing deaths, birth defects, and mental problems. Now, many beaches are closed occasionally or permanently due to pollution, and at a time when populations of fish have decreased, many areas are unsafe for fishing. Water pollution represents an especially dangerous problem in developing nations, which have high populations and manufacturing facilities that do not meet safety standards.

The most dangerous forms of water pollutants include sewage, which frequently contains dangerous pathogenic organisms; oil and hydrocarbons; heavy metals; radioactive substances; pesticides and herbicides; and corrosive substances such as acids and bases.

In developed countries, few direct sources of water pollution should exist, but homeowners still discharge motor oil, antifreeze, pet waste, and paint into storm sewers, and small manufacturers sometimes ignore proper disposal procedures. In developing countries, businesses and households often discharge wastes directly into streams or ponds that are also used for water supplies. Many sources contaminate water supplies indirectly. Indirect sources of pollution include runoff of waste from feedlots or runoff of agricultural chemicals from farmlands; leaking oil from pipelines, wells, or platforms; and large amounts of sediment from streets and parking lots.

Most industrial operations are required to treat wastewater before discharging it into rivers. Wastes from feedlots are collected in lagoons, settled, collected, and used for fertilizer. Heavy metals and organic compounds from industry are often reclaimed from wastewater and recycled, decreasing manufacturing costs. Sewage from homes undergoes at least two stages of treatment. Primary treatment consists of sedimentation and dyeing of solids, which may be used as fertilizer. Secondary treatment consists of aeration of the remaining liquid, through a process of stirring, trickling over filters, and spraying; aerobic bacteria oxidize much of the remaining organic matter. Tertiary treatment, using antibacterial agents such as chlorine or ozone, may be used to produce effluent water that is safe for further use.

Arsenic Contamination of Groundwater

Arsenic contamination of groundwater is a form of groundwater pollution which is often due to naturally occurring high concentrations of arsenic in deeper levels of groundwater. It is a high-profile problem due to the use of deep tubewells for water supply in the Ganges Delta, causing serious arsenic poisoning to large numbers of people. A 2007 study found that over 137 million people in more than 70 countries are probably affected by arsenic poisoning of drinking water. The problem became serious health concern after mass poisoning of water in Bangladesh. Arsenic contamination of ground water is found in many countries throughout the world, including the US.

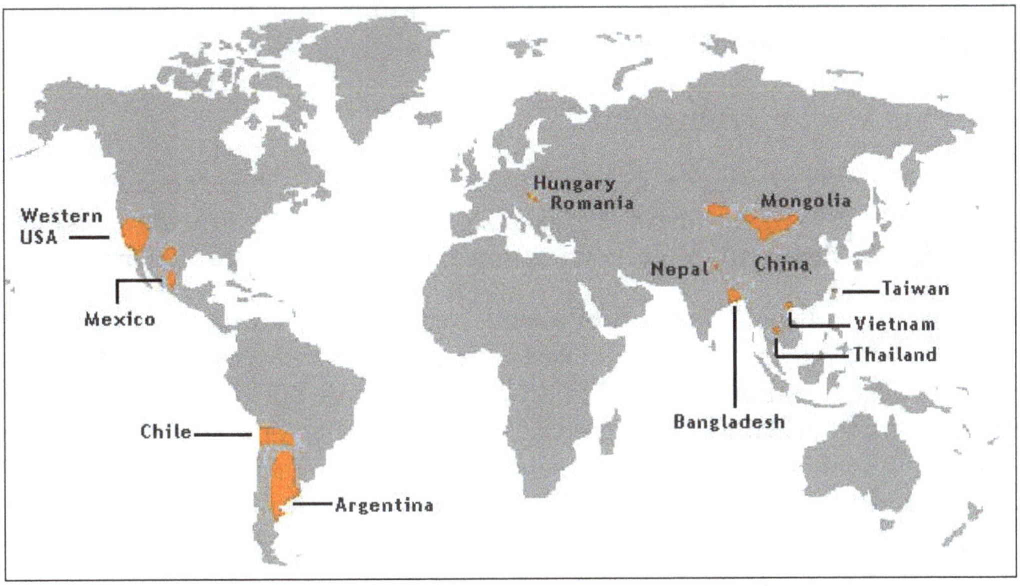

Groundwater arsenic contamination areas.

Approximately 20 major incidents of groundwater floarsenic contamination have been reported. Of these, four major incidents occurred in Asia, in Thailand, Taiwan, and Mainland China. Locations of potentially hazardous wells have been mapped in China.

Speciation of Arsenic Compounds in Water

Arsenic contaminated water typically contains arsenous acid and arsenic acid or their derivatives. Their names as "acids" is a formality; these species are not aggressive acids but are merely the soluble forms of arsenic near neutral pH. These compounds are extracted from the underlying rocks that surround the aquifer. Arsenic acid tends to exist as the ions $[HAsO_4]^{2-}$ and $[H_2AsO_4]^-$ in neutral water, whereas arsenous acid is not ionized.

Arsenic acid (H_3AsO_4), arsenous acid (H_3AsO_3) and their derivatives are typically encountered in arsenic contaminated ground water.

Water Purification Solutions

Access to clean drinking water is fraught with political, socio-economic, and cultural inequities. In practice, many water treatment strategies tend to be temporary fixes to a larger problem, often prolonging the social issues while treating the scientific ones. Scientific studies have shown that interdisciplinary approaches to water purification are especially important to consider, and long-lasting improvements involve larger perspectives than strict scientific approaches.

Small-scale Water Treatment

A review of methods to remove arsenic from groundwater in Pakistan summarizes the most technically viable inexpensive methods. Most small-scale treatments focus on water after it has left the distribution site, and are thus more focused on quick, temporary fixes.

A simpler and less expensive form of arsenic removal is known as the Sono arsenic filter, using three pitchers containing cast iron turnings and sand in the first pitcher and wood activated carbon and sand in the second. Plastic buckets can also be used as filter containers. It is claimed that thousands of these systems are in use and can last for years while avoiding the toxic waste disposal problem inherent to conventional arsenic removal plants. Although novel, this filter has not been certified by any sanitary standards such as NSF, ANSI, WQA and does not avoid toxic waste disposal similar to any other iron removal process.

In the United States small "under the sink" units have been used to remove arsenic from drinking water. This option is called "point of use" treatment. The most common types of domestic treatment use the technologies of adsorption (using media such as Bayoxide E33, GFH, activated alumina or titanium dioxide) or reverse osmosis. Ion exchange and activated alumina have been considered but not commonly used.

Chaff-based filters have been reported to reduce the arsenic content of water to 3 µg/L (3 ppb). This is especially important in areas where the potable water is provided by filtering the water extracted from the underground aquifer.

In iron electrocoagulation (Fe-EC), iron is dissolved nonstop using electricity, and the resulting ferric hydroxides, oxyhydoxides, and oxides form an absorbent readily attracted to arsenic. Current density, the amount of charge delivered per liter of water, of the process is often manipulated in order to achieve maximum arsenic depletion. This treatment strategy has primarily been used in Bangladesh, and has proven to be largely successful. In fact, using iron electrocoagulation to remove arsenic in water proved to be the most effective treatment option.

Large-scale Water Treatment

In some places, such as the United States, all the water supplied to residences by utilities must meet primary (health-based) drinking water standards. Regulations may need large-scale treatment systems to remove arsenic from the water supply. The effectiveness of any method depends on the chemical makeup of a particular water supply. The aqueous chemistry of arsenic is complex, and may affect the removal rate that can be achieved by a particular process.

Some large utilities with multiple water supply wells could shut down those wells with high arsenic concentrations, and produce only from wells or surface water sources that meet the arsenic standard. Other utilities, however, especially small utilities with only a few wells, may have no available water supply that meets the arsenic standard.

Coagulation/filtration (also known as flocculation) removes arsenic by coprecipitation and adsorption using iron coagulants. Coagulation/filtration using alum is already used by some utilities to remove suspended solids and may be adjusted to remove arsenic.

Iron oxide adsorption filters the water through a granular medium containing ferric oxide. Ferric oxide has a high affinity for adsorbing dissolved metals such as arsenic. The iron oxide medium eventually becomes saturated, and must be replaced. The sludge disposal is a problem here too.

Activated alumina is an adsorbent that effectively removes arsenic. Activated alumina columns connected to shallow tube wells in India and Bangladesh have removed both As(III) and As(V) from groundwater for decades. Long-term column performance has been possible through the efforts of community-elected water committees that collect a local water tax for funding operations and maintenance. It has also been used to remove undesirably high concentrations of fluoride.

Ion exchange has long been used as a water softening process, although usually on a single-home basis. Traditional anion exchange resins are effective in removing As(V), but not As(III), or arsenic trioxide, which doesn't have a net charge. Effective long-term ion exchange removal of arsenic requires a trained operator to maintain the column.

Both reverse osmosis and electrodialysis (also called *electrodialysis reversal*) can remove arsenic with a net ionic charge. (Note that arsenic oxide, As_2O_3, is a common form of arsenic in groundwater that is soluble, but has no net charge.) Some utilities presently use one of these methods to reduce total dissolved solids and therefore improve taste. A problem with both methods is the production of high-salinity waste water, called brine, or concentrate, which then must be disposed of.

Subterranean Arsenic Removal (SAR) Technology SAR Technology

In subterranean arsenic removal, aerated groundwater is recharged back into the aquifer to create an oxidation zone which can trap iron and arsenic on the soil particles through adsorption process. The oxidation zone created by aerated water boosts the activity of the arsenic-oxidizing microorganisms which can oxidize arsenic from +3 to +5 state SAR Technology. No chemicals are used and almost no sludge is produced during operational stage since iron and arsenic compounds are rendered inactive in the aquifer itself. Thus toxic waste disposal and the risk of its future mobilization is prevented. Also, it has very long operational life, similar to the long lasting tube wells drawing water from the shallow aquifers.

Six such SAR plants, funded by the World Bank and constructed by Ramakrishna Vivekananda Mission, Barrackpore & Queen's University Belfast, UK are operating in West Bengal. Each plant has been delivering more than 3,000 litres of arsenic and iron-free water daily to the rural community. The first community water treatment plant based on SAR technology was set up at Kashimpore near Kolkata in 2004 by a team of European and Indian engineers led by Bhaskar Sen Gupta of Queen's University Belfast for TiPOT.

SAR technology had been awarded Dhirubhai Ambani Award, 2010 from IChemE UK for Chemical Innovation. Again, SAR was the winner of the St. Andrews Award for Environment, 2010. The SAR Project was selected by the Blacksmith Institute – New York & Green Cross- Switzerland as one of the "12 Cases of Cleanup & Success" in the World's Worst Polluted Places Report 2009.

Currently, large scale SAR plants are being installed in US, Malaysia, Cambodia, and Vietnam.

Dietary Intake

Researchers from Bangladesh and the United Kingdom have claimed that dietary intake of arsenic adds a significant amount to total intake where contaminated water is used for irrigation.

VOC Contamination of Groundwater

VOCs (volatile organic compounds) are a dangerous contaminant of groundwater, leading to groundwater pollution. They are generally introduced to the environment through careless industrial practices.

VOCs are responsible for a number of adverse health effects especially for nursing and pregnant mothers. Many of these compounds were not known to be harmful until the late 1960s and it was some time before regular testing of groundwater identified these substances in drinking water sources.

Specific VOCs

PCE

Tetrachloroethylene is used in metal degreasing and dry cleaning. It is suspected to cause cancer, and a host of other health problems.

TCE

Trichloroethylene is used in cleaning metal parts and was widely believed to be harmless before being linked to birth defects. It is suspected to cause cancer, and a host of other health problems.

Toluene

Toluene is an organic compound which is mostly harmless to adults and is sometimes abused as an inhalant. Fetal toluene syndrome has been defined and resembles fetal alcohol syndrome with resultant birth defects, but the U.S. Centers for Disease Control and Prevention have identified differentiating features between the two syndromes including FTS having the additional facial features of micrognathia, large anterior fontanel, down-turned mouth corners, hair patterning abnormalities, bifrontal narrowing of the face, and ear abnormalities.

Biochemical Treatment of Wastewater

Anaerobic Process

Anaerobic methods involve treatment of wastewater without access to oxygen. Instead, methane fermentation is used. The advantage of this method is the high level of converting contaminants with the formation of biogas byproduct.

Wastewater contaminated by nitrogen compounds are converted in the process of denitrification by microbes (such as Paracoccus denitrsficans) with formation of gaseous oxide (NO_2) and molecular nitrogen (N_2) or ammonia (NH_3).

The fermentation occurs in sealed methane tanks (septic tanks). These are air tight reactors filled with immobile biological sludge, sludge ponds or similar. Methane tanks not only treat wastewaste, but also generate gas, which has high calorific value.

The main parameters of anaerobic methane fermentation are temperature, amount of waste and the intensity of mixing the water with the sludge. It should be noted that it is impossible to achieve full fermentation of organic substances in methane tanks, since the process is slow and requires stable and favorable temperature conditions. The degree of organic decomposition in the result is 40% (with methane output of 70%). An important point about the anaerobic process is the slight increase of microbial mass (by an order of magnitude less than in aerobic conditions), it also does not require the removal of large amounts of biological sludge from the reactor.

Biochemical Treatment of Wastewater: The products of organic destruction, which are formed in the first trophic level, function as substrate for second level microbes. The characteristics of the intermediary products of anaerobic fermentation depend on the composition of the initial contaminants.

Anaerobic oxidation, as a rule, is efficient for treatment of wastewater contaminated with large amounts of organic substances (it ensures decomposition of over a hundred organic compounds).

Biochemical Oxygen Demand

Biochemical Oxygen Demand (BOD) is the amount of dissolved oxygen needed by aerobic biological organisms to break down organic material.

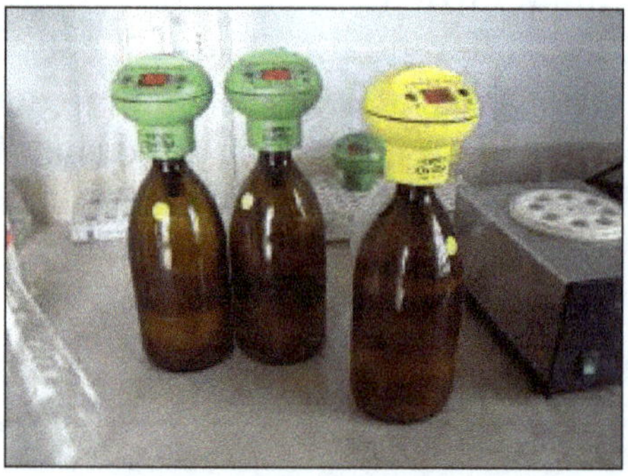

BOD test bottles at the laboratory of a wastewater treatment plant.

Biochemical oxygen demand (BOD) is the amount of dissolved oxygen needed (i.e. demanded) by aerobic biological organisms to break down organic material present in a given water sample at certain temperature over a specific time period. The BOD value is most commonly expressed in milligrams of oxygen consumed per litre of sample during 5 days of incubation at 20 °C and is often used as a surrogate of the degree of organic pollution of water.

BOD reduction is used as a gauge of the effectiveness of wastewater treatment plants. BOD of wastewater effluents is used to indicate the short term impact on the oxygen levels of the receiving water.

BOD analysis is similar in function to chemical oxygen demand (COD) analysis, in that both measure the amount of organic compounds in water. However, COD analysis is less specific, since it measures everything that can be chemically oxidized, rather than just levels of biologically oxidized organic matter.

Most natural waters contain small quantities of organic compounds. Aquatic microorganisms have evolved to use some of these compounds as food. Microorganisms living in oxygenated waters use dissolved oxygen to oxidatively degrade the organic compounds, releasing energy which is used for growth and reproduction. Populations of these microorganisms tend to increase in proportion to the amount of food available. This microbial metabolism creates an oxygen demand proportional to the amount of organic compounds useful as food. Under some circumstances, microbial metabolism can consume dissolved oxygen faster than atmospheric oxygen can dissolve into the water or the autotrophic community (algae, cyanobacteria and macrophytes) can produce. Fish and aquatic insects may die when oxygen is depleted by microbial metabolism.

Biochemical oxygen demand is the amount of oxygen required for microbial metabolism of organic compounds in water. This demand occurs over some variable period of time depending on temperature, nutrient concentrations, and the enzymes available to indigenous microbial populations. The amount of oxygen required to completely oxidize the organic compounds to carbon dioxide and water through generations of microbial growth, death, decay, and cannibalism is total biochemical oxygen demand (total BOD). Total BOD is of more significance to food webs than to water quality. Dissolved oxygen depletion is most likely to become evident during the initial aquatic microbial population explosion in response to a large amount of organic material. If the microbial population deoxygenates the water, however, that lack of oxygen imposes a limit on population growth of aerobic aquatic microbial organisms resulting in a longer term food surplus and oxygen deficit.

A standard temperature at which BOD testing should be carried out was first proposed by the Royal Commission on Sewage Disposal in its eighth report in 1912:

> "An effluent in order to comply with the general standard must not contain as discharged more than 3 parts per 100,000 of suspended matter, and with its suspended matters included must not take up at 65 °F more than 2.0 parts per 100,000 of dissolved oxygen in 5 days. This general standard should be prescribed either by Statute or by order of the Central Authority, and should be subject to modifications by that Authority after an interval of not less than ten years."

This was later standardised at 68 °F and then 20 °C. This temperature may be significantly different from the temperature of the natural environment of the water being tested.

Although the Royal Commission on Sewage Disposal proposed 5 days as an adequate test period for rivers of the United Kingdom of Great Britain and Ireland, longer periods were investigated for North American rivers. Incubation periods of 1, 2, 5, 10 and 20 days were being used into the mid-20th century. Keeping dissolved oxygen available at their chosen temperature, investigators found up to 99 percent of total BOD was exerted within 20 days, 90 percent within 10 days, and approximately 68 percent within 5 days. Variable microbial population shifts to nitrifying bacteria limit test reproducibility for periods greater than 5 days. The 5-day test protocol with acceptably reproducible results emphasizing carbonaceous BOD has been endorsed by the United States Environmental Protection Agency (EPA). This 5-day BOD test result may be described as the amount of oxygen required for aquatic microorganisms to stabilize decomposable organic matter under aerobic conditions. Stabilization, in this context, may be perceived in general terms as the conversion of food to living aquatic fauna. Although these fauna will continue to exert biochemical oxygen demand as they die, that tends to occur within a more stable evolved ecosystem including higher trophic levels.

Taking samples from the influent raw wastewater stream for BOD measurements at a wastewater treatment plant in Haran-Al-Awamied near Damascus in Syria.

Use in Sewage Treatment

The BOD is used in measuring waste loadings to treatment plants and in evaluating the BOD-removal efficiency of such treatment systems.

Methods

Winkler published the methodolgy of a simple, accurate and direct dissolved oxygen analytical procedure in 1888. Since that time, the analysis of dissolved oxygen levels for water has been key to the determination of surface water. The Winkler method is still one of only two analytical techniques used to calibrate oxygen electrode meters; the other procedure is based on oxygen solubility at saturation as per Henry's law. There are two recognized methods for the measurement of dissolved oxygen for BOD and a number of other methods not currently internationally recognised as standard methods.

Dilution Method

Disposable BOD bottle.

Glass BOD bottle.

This standard method is recognized by EPA, which is labeled Method 5210B in the *Standard Methods for the Examination of Water and Wastewater*. In order to obtain BOD_5, dissolved oxygen (DO) concentrations in a sample must be measured before and after the incubation period, and appropriately adjusted by the sample corresponding dilution factor. This analysis is performed using 300 ml incubation bottles in which buffered dilution water is dosed with seed microorganisms and stored for 5 days in the dark room at 20 °C to prevent DO production via photosynthesis. The bottles have traditionally been made of glass, which required cleaning and rinsing between samples. A SM 5210B approved, disposable, plastic BOD bottle is available which eliminates this step. In addition to the various dilutions of BOD samples, this procedure requires dilution water blanks, glucose glutamic acid (GGA) controls, and seed controls.

The dilution water blank is used to confirm the quality of the dilution water that is used to dilute the other samples. This is necessary because impurities in the dilution water may cause significant alterations in the results. The GGA control is a standardized solution to determine the quality of the seed, where its recommended BOD_5 concentration is 198 mg/l ± 30.5 mg/l. For measurement of carbonaceous BOD (cBOD), a nitrification inhibitor is added after the dilution water has been added to the sample. The inhibitor hinders the oxidation of ammonia nitrogen, which supplies the nitrogenous BOD (nBOD). When performing the BOD_5 test, it is conventional practice to measure only cBOD because nitrogenous demand does not reflect the oxygen demand from organic matter. This is because nBOD is generated by the breakdown of proteins, whereas cBOD is produced by the breakdown of organic molecules.

BOD_5 is calculated by:

- Unseeded: $BOD_5 = \frac{(D_o - D_5)}{P}$

- Seeded: $BOD_5 = \frac{(D_0 - D_5) - (B_0 - B_5)f}{P}$

where:

D_0 is the dissolved oxygen (D_0) of the diluted solution after preparation (mg/l).

D_5 is the DO of the diluted solution after 5 day incubation (mg/l).

P is the decimal dilution factor.

B_0 is the DO of diluted seed sample after preparation (mg/l).

B_5 is the DO of diluted seed sample after 5 day incubation (mg/l).

f is the ratio of seed volume in dilution solution to seed volume in BOD test on seed.

Manometric Method

This method is limited to the measurement of the oxygen consumption due only to carbonaceous oxidation. Ammonia oxidation is inhibited.

The sample is kept in a sealed container fitted with a pressure sensor. A substance that absorbs carbon dioxide (typically lithium hydroxide) is added in the container above the sample level. The sample is stored in conditions identical to the dilution method. Oxygen is consumed and, as ammonia oxidation is inhibited, carbon dioxide is released. The total amount of gas, and thus the pressure, decreases because carbon dioxide is absorbed. From the drop of pressure, the sensor electronics computes and displays the consumed quantity of oxygen.

The main advantages of this method compared to the dilution method are:

- Simplicity: No dilution of sample required, no seeding, no blank sample.
- Direct reading of BOD value.
- Continuous display of BOD value at the current incubation time.

Alternative Methods

Biosensor

An alternative to measure BOD is the development of biosensors, which are devices for the detection of an analyte that combines a biological component with a physicochemical detector component. Enzymes are the most widely used biological sensing elements in the fabrication of biosensors. Their application in biosensor construction is limited by the tedious, time consuming and costly enzyme purification methods. Microorganisms provide an ideal alternative to these bottlenecks.

Many micro organisms useful for BOD assessment are relatively easy to maintain in pure cultures, grow and harvest at low cost. Moreover, the use of microbes in the field of biosensors has opened up new possibilities and advantages such as ease of handling, preparation and low cost of device. A number of pure cultures, e.g. *Trichosporon cutaneum, Bacillus cereus, Klebsiella oxytoca, Pseudomonas sp.* etc. individually, have been used by many workers for the construction of BOD biosensor. On the other hand, many workers have immobilized activated sludge, or a mixture of two or three bacterial species and on various membranes for the construction of BOD biosensor. The most commonly used membranes were polyvinyl alcohol, porous hydrophilic membranes etc.

A defined microbial consortium can be formed by conducting a systematic study, i.e. pre-testing of selected micro-organisms for use as a seeding material in BOD analysis of a wide variety of industrial effluents. Such a formulated consortium can be immobilized on suitable membrane, i.e. charged nylon membrane. Charged nylon membrane is suitable for microbial immobilization, due to the specific binding between negatively charged bacterial cell and positively charged nylon membrane. So the advantages of the nylon membrane over the other membranes are: The dual binding, i.e. Adsorption as well as entrapment, thus resulting in a more stable immobilized membrane. Such specific Microbial consortium based BOD analytical devices, may find great application in monitoring of the degree of pollutant strength, in a wide variety of industrial waste water within a very short time.

Biosensors can be used to indirectly measure BOD via a fast (usually <30 min) to be determined BOD substitute and a corresponding calibration curve method. Consequently, biosensors are now commercially available, but they do have several limitations such as their high maintenance costs, limited run lengths due to the need for reactivation, and the inability to respond to changing quality characteristics as would normally occur in wastewater treatment streams; e.g. diffusion processes of the biodegradable organic matter into the membrane and different responses by different microbial species which lead to problems with the reproducibility of result. Another important limitation is the uncertainty associated with the calibration function for translating the BOD substitute into the real BOD.

Fluorescent

A surrogate to BOD_5 has been developed using a resazurin derivative which reveals the extent of oxygen uptake by micro-organisms for organic matter mineralization. A cross-validation performed on 109 samples in Europe and the United-States showed a strict statistical equivalence between results from both methods. An electrode whas been developed based on the luminescence emission of a photo-active chemical compound and the quenching of that emission by oxygen. This quenching photophysics mechanism is described by the Stern–Volmer equation for dissolved oxygen in a solution:

$$I_o / I = 1 + K_{SV} [O_2].$$

- I : Luminescence in the presence of oxygen.
- I_o : Luminescence in the absence of oxygen.
- K_{SV} : Stern-Volmer constant for oxygen quenching.
- $[O_2]$: Dissolved oxygen concentration.

The determination of oxygen concentration by luminescence quenching has a linear response over a broad range of oxygen concentrations and has excellent accuracy and reproducibility.

Polarographic Method

The development of an analytical instrument that utilizes the reduction-oxidation (redox) chemistry of oxygen in the presence of dissimilar metal electrodes was introduced during the 1950s. This redox electrode utilized an oxygen-permeable membrane to allow the diffusion of the gas into an electrochemical cell and its concentration determined by polarographic or galvanic electrodes. This analytical method is sensitive and accurate down to levels of ± 0.1 mg/l dissolved oxygen. Calibration of the redox electrode of this membrane electrode still requires the use of the Henry's law table or the Winkler test for dissolved oxygen.

Software Sensor

Rustum et al. proposed the use of the KSOM to develop intelligent models for making rapid inferences about BOD using other easy to measure water quality parameters, which, unlike BOD, can be obtained directly and reliably using on-line hardware sensors. This will make the use of BOD for on-line process monitoring and control a more plausible proposition. In comparison to other data-driven modeling paradigms such as multi-layer perceptrons artificial neural networks (MLP ANN) and classical multi-variate regression analysis, the KSOM is not negatively affected by missing data. Moreover, time sequencing of data is not a problem when compared to classical time series analysis.

Real-time BOD Monitoring

Until recently, real-time monitoring of BOD was unattainable owing to its complex nature. Recent research by a leading UK university has discovered the link between multiple water quality parameters including electrical conductivity, turbidity, TLF and CDOM. These parameters are all capable of being monitored in real-time through a combination of traditional methods (electrical conductivity via electrodes) and newer methods such as fluorescence. The monitoring of tryptophan-like fluorescence (TLF) has been successfully utilised as a proxy for biological activity and enumeration, particularly with a focus on *Escherichia coli* (E. Coli). TLF based monitoring is applicable across a wide range of environments, including but by no means limited to sewage treatment works and freshwaters. Therefore, there has been a significant movement towards combined sensor systems that can monitor parameters and use them, in real-time, to provide a reading of BOD that is of laboratory quality.

Dissolved Oxygen Probes: Membrane and Luminescence

The development of an analytical instrument that utilizes the reduction-oxidation (redox) chemistry of oxygen in the presence of dissimilar metal electrodes was introduced during the 1950s. This redox electrode (also known as dissolved oxygen sensor) utilized an oxygen-permeable membrane to allow the diffusion of the gas into an electrochemical cell and its concentration determined by polarographic or galvanic electrodes. This analytical method is sensitive and accurate to down to levels of ± 0.1 mg/l dissolved oxygen. Calibration of the redox electrode of this membrane electrode still requires the use of the Henry's law table or the Winkler test for dissolved oxygen.

Test Limitations

The test method involves variables limiting reproducibility. Tests normally show observations varying plus or minus ten to twenty percent around the mean.

Toxicity

Some wastes contain chemicals capable of suppressing microbiological growth or activity. Potential sources include industrial wastes, antibiotics in pharmaceutical or medical wastes, sanitizers in food processing or commercial cleaning facilities, chlorination disinfection used following conventional sewage treatment, and odor-control formulations used in sanitary waste holding tanks in passenger vehicles or portable toilets. Suppression of the microbial community oxidizing the waste will lower the test result.

Appropriate Microbial Population

The test relies upon a microbial ecosystem with enzymes capable of oxidizing the available organic material. Some waste waters, such as those from biological secondary sewage treatment, will already contain a large population of microorganisms acclimated to the water being tested. An appreciable portion of the waste may be utilized during the holding period prior to commencement of the test procedure. On the other hand, organic wastes from industrial sources may require specialized enzymes. Microbial populations from standard seed sources may take some time to produce those enzymes. A specialized seed culture may be appropriate to reflect conditions of an evolved ecosystem in the receiving waters.

Chemical Wastewater Treatment

Chemicals are used during wastewater treatment in an array of processes to expedite disinfection. These chemical processes, which induce chemical reactions, are called chemical unit processes and are used alongside biological and physical cleaning processes to achieve various water standards. Specialized chemicals such as chlorine, hydrogen peroxide, sodium chlorite, and sodium hypochlorite (bleach) act as agents that disinfect, sanitize, and assist in the purification of wastewater at treatment facilities. There are several distinct chemical unit processes, including chemical coagulation, chemical precipitation, chemical oxidation, and advanced oxidation, ion exchange, and chemical neutralization and stabilization, which can be applied to wastewater during cleaning.

Processes of Chemical Wastewater Treatment

Chemical Precipitation

Chemical precipitation is the most common method for removing dissolved metals from wastewater solution containing toxic metals. To convert the dissolved metals into solid particle form, a precipitation reagent is added to the mixture. A chemical reaction, triggered by the reagent, causes the dissolved metals to form solid particles. Filtration can then be used to remove the particles from the mixture. How well the process works is dependent upon the kind of metal present, the concentration of the metal, and the kind of reagent used. In hydroxide precipitation, a commonly used chemical precipitation process, calcium or sodium hydroxide is used as the reagent to create solid metal hydroxides. However, it can be difficult to create hydroxides from dissolved metal particles in wastewater because many wastewater solutions contain mixed metals.

Chemical Coagulation

This chemical process involves destabilizing wastewater particles so that they aggregate during chemical flocculation. Fine solid particles dispersed in wastewater carry negative electric surface charges (in their normal stable state), which prevent them from forming larger groups and settling. Chemical coagulation destabilizes these particles by introducing positively charged coagulants that then reduce the negative particles' charge. Once the charge is reduced, the particles freely form larger groups. Next, an anionic flocculant is introduced to the mixture. Because the flocculant reacts against the positively charged mixture, it either neutralizes the particle groups or creates bridges between them to bind the particles into larger groups. After larger particle groups are formed, sedimentation can be used to remove the particles from the mixture.

Chemical Oxidation and Advanced Oxidation

With the introduction of an oxidizing agent during chemical oxidation, electrons move from the oxidant to the pollutants in wastewater. The pollutants then undergo structural modification, becoming less destructive compounds. Alkaline chlorination uses chlorine as an oxidant against cyanide. However, alkaline chlorination as a chemical oxidation process can lead to the creation of toxic chlorinated compounds, and additional steps may be required. Advanced oxidation can help remove any organic compounds that are produced as a byproduct of chemical oxidation, through processes such as steam stripping, air stripping, or activated carbon adsorption.

Ion Exchange

When water is too hard, it is difficult to use to clean and often leaves a grey residue. (This is why clothing washed in hard water often retains a dingy tint.) An ion exchange process, similar to the reverse osmosis process, can be used to soften the water. Calcium and magnesium are common ions that lead to water hardness. To soften the water, positively charged sodium ions are introduced in the form of dissolved sodium chloride salt or brine. Hard calcium and magnesium ions exchange places with sodium ions, and free sodium ions are simply released in the water. However, after softening a large amount of water, the softening solution may fill with excess calcium and magnesium ions, requiring the solution to be recharged with sodium ions.

Chemical Stabilization

This chemical wastewater treatment process works in a similar fashion as chemical oxidation. Sludge is treated with a large amount of a given oxidant, such as chlorine. The introduction of the oxidant slows down the rate of biological growth within the sludge and also helps deodorize the mixture. The water is then removed from the sludge. Hydrogen peroxide can also be used as an oxidant and may be a more cost-effective choice.

References

- Atmospheric-chemistry, earth-and-planetary-sciences: sciencedirect.com, Retrieved 6 April, 2020
- Törnqvist, H. K.; Mills, N. L.; Gonzalez, M.; Miller, M. R.; Robinson, S. D.; Megson, I. L.; MacNee, W.; Donaldson, K.; Söderberg, S.; Newby, D. E.; Sandström, T.; Blomberg, A. (2007). "Persistent Endothelial Dysfunction in Humans after Diesel Exhaust Inhalation". American Journal of Respiratory and Critical Care Medicine. 176 (4): 395–400. doi:10.1164/rccm.200606-872OC. PMID 17446340
- Introduction-water-chemistry, chemical, water-quality: freedrinkingwater.com, Retrieved 19 April, 2020
- "Investigation of pollution incidents". Queensland Government - Department of Environment and Heritage Proetection. 21 July 2016. Retrieved 1 August 2016
- Hering Janet G.; Katsoyiannis Ioannis A.; Theoduloz Gerardo Ahumada; Berg Michael; Hug Stephan J. (2017-05-01). "Arsenic Removal from Drinking Water: Experiences with Technologies and Constraints in Practice". Journal of Environmental Engineering. 143 (5): 03117002. doi:10.1061/(ASCE)EE.1943-7870.0001225
- Wastewater-chemical-treatment, chemicals: thomasnet.com, Retrieved 16 May, 2020
- Clair N. Sawyer; Perry L. McCarty; Gene F. Parkin (2003). Chemistry for Environmental Engineering and Science (5th ed.). New York: McGraw-Hill. ISBN 978-0-07-248066-5

3

Soil Biochemistry

Soil biochemistry is a branch of soil science that is associated with the formation and decomposition of soil organic matter, biochemical reactions of carbon, nitrogen, phosphorus, sulfur, metals and xenobiotics in soils, etc. This chapter closely examines the aspects of soil biochemistry to provide an extensive understanding of soil biochemistry.

Soil

Soil is the biologically active, porous medium that has developed in the uppermost layer of Earth's crust. Soil is one of the principal substrata of life on Earth, serving as a reservoir of water and nutrients, as a medium for the filtration and breakdown of injurious wastes, and as a participant in the cycling of carbon and other elements through the global ecosystem. It has evolved through weathering processes driven by biological, climatic, geologic, and topographic influences.

Chernozem soil profile from Germany, showing a thick humus-rich surface horizon with a light-coloured lime-rich layer below.

Since the rise of agriculture and forestry in the 8th millennium BCE, there has also arisen by necessity a practical awareness of soils and their management. In the 18th and 19th centuries the Industrial Revolution brought increasing pressure on soil to produce raw materials demanded by commerce, while the development of quantitative science offered new opportunities for improved soil management. The study of soil as a separate scientific discipline began about the same time with systematic investigations of substances that enhance plant growth. This initial inquiry has expanded to an understanding of soils as complex, dynamic, biogeochemical systems that are vital to the life cycles of terrestrial vegetation and soil-inhabiting organisms—and by extension to the human race as well.

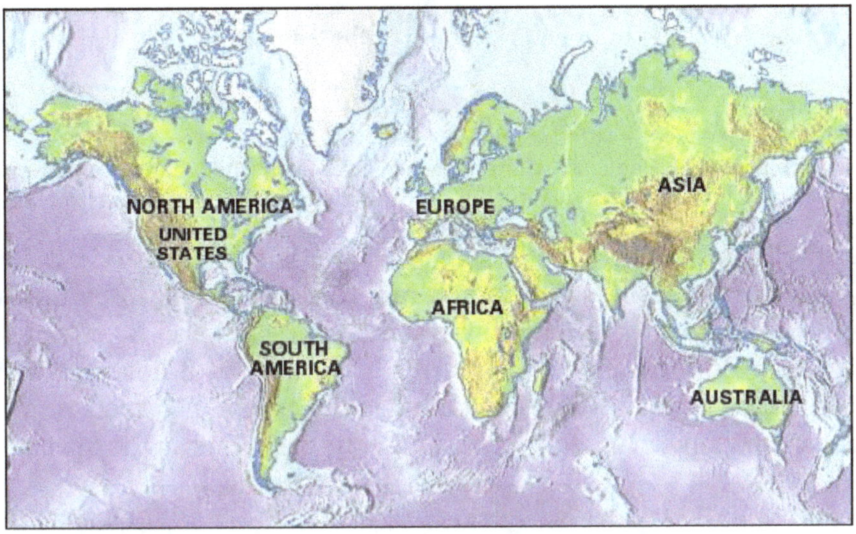

Interactive soil map of the world.

The Soil Profile

Soil Horizons

Soils differ widely in their properties because of geologic and climatic variation over distance and time. Even a simple property, such as the soil thickness, can range from a few centimetres to many metres, depending on the intensity and duration of weathering, episodes of soil deposition and erosion, and the patterns of landscape evolution. Nevertheless, in spite of this variability, soils have a unique structural characteristic that distinguishes them from mere earth materials and serves as a basis for their classification: A vertical sequence of layers produced by the combined actions of percolating waters and living organisms.

These layers are called horizons, and the full vertical sequence of horizons constitutes the soil profile. Soil horizons are defined by features that reflect soil-forming processes. For instance, the uppermost soil layer (not including surface litter) is termed the A horizon. This is a weathered layer that contains an accumulation of humus (decomposed, dark-coloured, carbon-rich matter) and microbial biomass that is mixed with small-grained minerals to form aggregate structures.

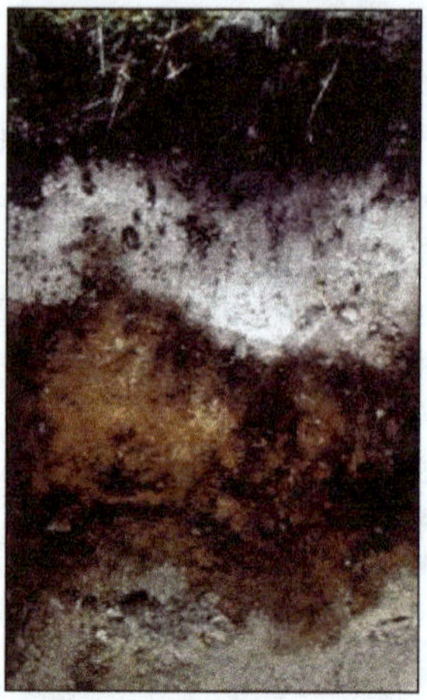

Podzol soil profile from Ireland, showing a bleached layer from which humus and metal oxides have been leached and subsequently deposited in the typically reddish horizon below.

Below A lies the B horizon: In mature soils this layer is characterized by an accumulation of clay (small particles less than 0.002 mm [0.00008 inch] in diameter) that has either been deposited out of percolating waters or precipitated by chemical processes involving dissolved products of weathering. Clay endows B horizons with an array of diverse structural features (blocks, columns, and prisms) formed from small clay particles that can be linked together in various configurations as the horizon evolves.

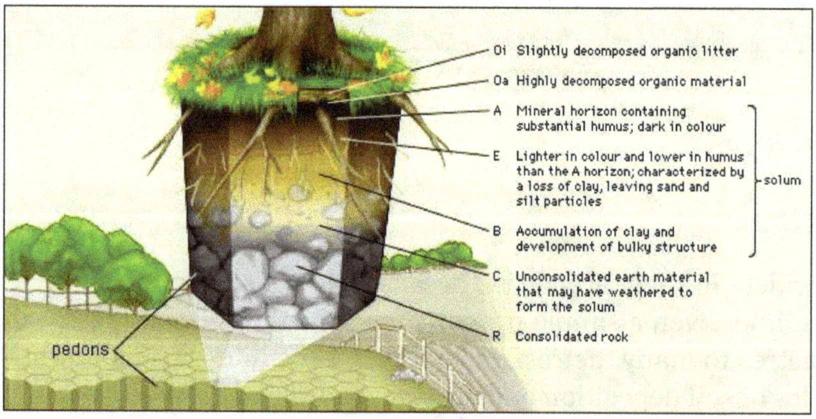

The soil profile, showing the major layers from the O horizon (organic material) to the R horizon (consolidated rock). A pedon is the smallest unit of land surface that can be used to study the characteristic soil profile of a landscape.

Below the A and B horizons is the C horizon, a zone of little or no humus accumulation or soil structure development. The C horizon often is composed of unconsolidated parent material from which the A and B horizons have formed. It lacks the characteristic features of the A and B horizons and may be either relatively unweathered or deeply weathered. At some depth below the A, B, and C horizons lies consolidated rock, which makes up the R horizon.

These simple letter designations are supplemented in two ways. First, two additional horizons are defined. Litter and decomposed organic matter (for example, plant and animal remains) that typically lie exposed on the land surface above the A horizon are given the designation O horizon, whereas the layer immediately below an A horizon that has been extensively leached (that is, slowly washed of certain contents by the action of percolating water) is given the separate designation E horizon, or zone of eluviation. The development of E horizons is favoured by high rainfall and sandy parent material, two factors that help to ensure extensive water percolation. The solid particles lost through leaching are deposited in the B horizon, which then can be regarded as a zone of illuviation. The combined A, E, B horizon sequence is called the solum. The solum is the true seat of soil-forming processes and is the principal habitat for soil organisms. (Transitional layers, having intermediate properties, are designated with the two letters of the adjacent horizons).

The second enhancement to soil horizon nomenclature is the use of lowercase suffixes to designate special features that are important to soil development. The most common of these suffixes are applied to B horizons: G to denote mottling caused by waterlogging, h to denote the illuvial accumulation of humus, k to denote carbonate mineral precipitates, o to denote residual metal oxides, s to denote the illuvial accumulation of metal oxides and humus, and t to denote the accumulation of clay.

Pedons and Polypedons

Soils are natural elements of weathered landscapes whose properties may vary spatially. For scientific study, however, it is useful to think of soils as unions of modules known as pedons. A pedon is the smallest element of landscape that can be called soil. Its depth limit is the somewhat arbitrary boundary between soil and "not soil" (e.g., bedrock). Its lateral dimensions must be large enough to permit a study of any horizons present—in general, an area from 1 to 10 square metres (10 to 100 square feet), taking into account that a horizon may be variable in thickness or even discontinuous. Wherever horizons are cyclic and recur at intervals of 2 to 7 metres (7 to 23 feet), the pedon includes one-half the cycle. Thus, each pedon includes the range of horizon variability that occurs within small areas. Wherever the cycle is less than 2 metres, or wherever all horizons are continuous and of uniform thickness, the pedon has an area of 1 square metre.

Soils are encountered on the landscape as groups of similar pedons, called polypedons, that contain sufficient area to qualify as a taxonomic unit. Polypedons are bounded from below by "not soil" and laterally by pedons of dissimilar characteristics.

Soil Chemistry

Soil chemistry can be considered as the natural chemical composition of a given soil. This natural chemical composition of a soil is a function of that soil's parent material.

In many areas of the world soil is formed in place and derived directly from the weathering and degradation of rocks. When soil is derived from rocks, its soil chemistry is a direct reflection of the rocks' chemistry, including the minerals found in the rocks. There are different forms of rocks, including the major classifications of sedimentary, metamorphic, and igneous. Within each of these major separations there are different categories of rocks. Thus, rocks are a collection of different minerals. Soils derived from these different rocks will have different chemical complexes. For example, soils derived from sandstone will have a limited array of different chemicals, whereas soils derived from igneous rock might have a greater diversity of minerals and chemicals. However, not all soils are formed in situ from the rocks that exist in place.

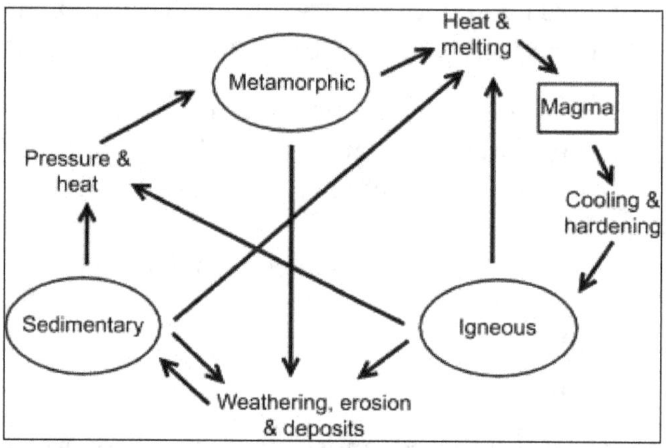

Simplified rock cycle and major rock material available to form soil. Rocks vary in their mineral and chemical composition, which in turn affects the type of soil derived from these rocks.

Soil parent materials range from glacial drift (glacial till), to water deposited (fluvial) material, and mass wasting of primary rock or other deposit caused by gravitational forces (colluvium), among others. Soil that is derived from glacial drift or any of these other sources will likely have a very complex chemical composition reflecting the many rock types and other materials entrained in the parent material. Thus, a natural soil's chemical composition is a reflection of its source materials. While soil will inherit a given mineral chemical composition, the organic composition is derived from living organisms (e.g., animals, plants, fungi). It is this organic matter that is often considered the key to good soil health.

Natural soil chemistry can be changed by various natural forces such as leaching of chemical elements by water moving through the soil, chemical reactions, and biological activity. However, soil chemistry can also be altered by human impact from various land uses, including farming. It is a given that soil chemistry can, and will, be changed by natural forces, and one might consider these changes negative or positive impacts depending on the intended future use of the soil. Similar to natural changes, management of soils by humans can result in positive or negative changes to soil chemistry, such as increases in soil organic matter or soil erosion because of aggressive tillage practices. For example, fertilizer and manure applications to a silt loam soil increased the percentage of water-stable aggregates of a silt loam in Romania because of increases in soil organic matter. Other management practices, such as inversion tillage, can increase the erodibility of soil because of soil aggregate disturbance.

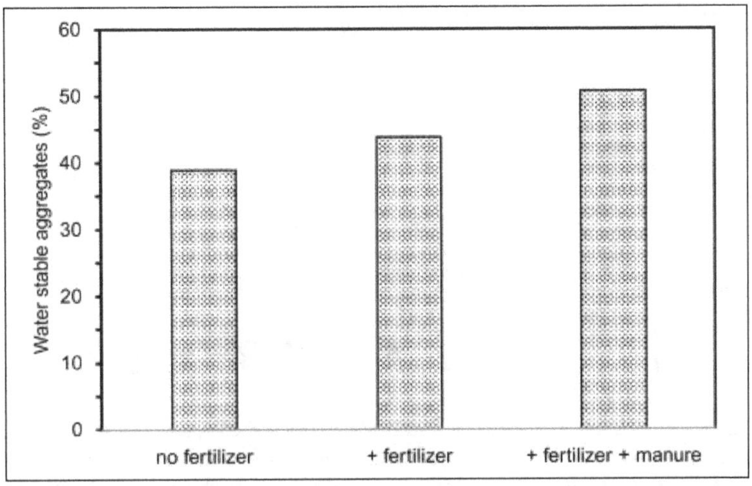

Impact of fertilizer and manure application to soybean-wheat-corn-barley rotation on water-stable aggregates of a silt loam from northeastern Romania.

The greatest negative effect of CAPS on soil chemistry is not only the intense use and introduction of herbicides and insecticides in the soil ecosystem, but the lack of crop rotations and cover crops which are known for adding organic carbon (C) and biological diversity to the soil system. The tillage systems used in most CAPS result in substantial soil disturbance and generally a negative impact on soil organic matter. Aggressively tilling the soil leads to a considerable incorporation of air, which causes a spike in microbial activity and subsequently accelerates soil organic matter breakdown. Since soil organic matter is a key to good soil health, any reductions in soil organic matter will obviously reduce soil health.

It has been shown that excessive introduction of certain N fertilizers into the soil ecosystem results in soil acidification. This added acidity further degrades soil by affecting soil primary particles, namely clay minerals. This was demonstrated by Barak et al. for glacially deposited soils in south central Wisconsin, United States, where they found a significant increase in soil acidity caused decreases in cation exchange capacity (CEC), base saturation, and exchangeable calcium (Ca) and magnesium (Mg). Some of the degradations in soil properties were considerable, e.g., clays in this soil showed a 16% loss of CEC. These degradations came from application of anhydrous ammonia as a N fertilizer in long-term fertility experiments. These results highlight the impact that CAPS management practices can have on soil health and productivity.

In addition to fertilizer applications, soil chemistry is altered by pesticide applications. When pesticides, especially insecticides are applied to the soil they sometimes change the soil chemical environment to the detriment of soil biological populations. Good soil quality is characterized as having a vast range of soil microorganisms. While soil-applied insecticides have been proven to generally kill target insects, they will also destroy "beneficial" organisms.

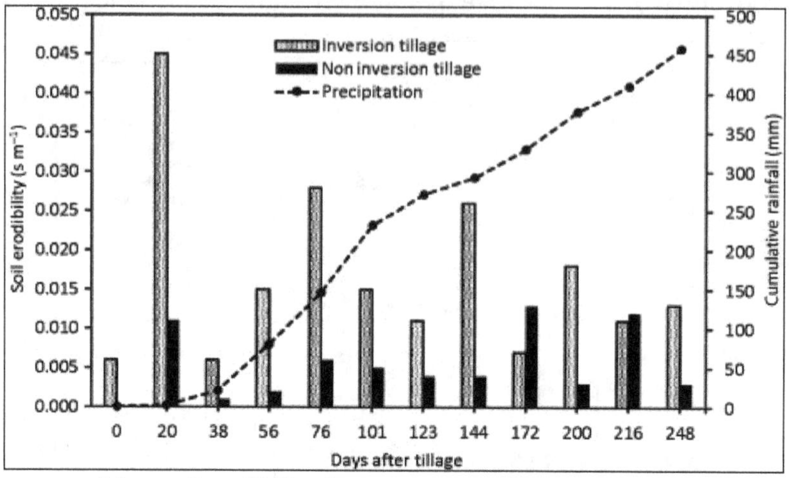

The concentrated flow soil erodibility factor increases after inversion tillage compared to noninversion tillage because of greater soil aggregate disturbance of a Belgian silt loam.

When excessive pesticides are used, which is generally the case in CAPS, not only do we see degradation of the soil ecosystem, but the underlying groundwater system often becomes polluted. Soil chemistry is altered by leaching of chemical substances by water. Chemical leaching is not limited to natural soil chemical elements but synthetic chemicals applied to soil, such as pesticides and fertilizers, leach as well. This is especially true for sandy soils and soils with little organic matter content. Most pesticides are strongly retained by soil organic matter and clay minerals. Sandy soils are often low in organic matter and CEC, thus providing ideal conditions for pesticides and nitrate from organic and synthetic N fertilizers to leach. In addition, sandy soils tend to be well drained, thus water can move rapidly through them. Groundwater contamination is often reported in areas where the water table is close to the soil surface.

Soil chemistry has traditionally focused on the chemical reactions in soils that affect plant growth and plant nutrition. However, beginning in the 1970s and certainly in the 1990s, as concerns increased about inorganic and organic contaminants in water and soil and their impact on plant, animal, and human health, the emphasis of soil chemistry is now on environmental soil chemistry. Environmental soil chemistry is the study of chemical reactions between soils and environmentally important plant nutrients, radionuclides, metals, metalloids, and organic chemicals.

Soil Biochemistry

Soil biochemistry is one of the branches of soil science dealing with the formation and decomposition of soil organic matter, biochemical reactions of carbon, nitrogen, phosphorus, sulfur, metals and xenobiotics in soils, and biochemistry of the plant-root rhizosphere. Soil can be considered as a tissue, containing mineralogical compounds, organic matter, water and air in which biochemical processes occur. These reactions govern elemental cycles, decomposition and synthesis of humic substances, formation of soil structure and behavior of a variety of pollutants. Soil biochemistry is one of the branches of soil science dealing with the formation and decomposition of soil organic matter, reactions of the carbon, nitrogen, phosphorus, and sulfur cycles, metal biochemistry of soils, and the biochemistry of the plant-root rhizosphere.

Soil biochemical reactions concern primarily the organic fraction, though it is made up of only 5% of the total soil volume. This fraction serves as a storehouse of essential nutrients for plant and microbial growth, is a major source of energy for microorganisms, and plays an important role in determining soil structure. The biochemical reactions associated with the organic matter fraction are greatly affected by shifts in environmental factors such as oxidation-reduction (redox) potentials, pH, water and temperature.

The redox potential (E_h) is the measure of a substance to donate or accept electrons. It is directly affected by oxygen concentration and is also correlated with pH. In soil only a narrow range of redox potentials are possible because of the instability of water in strong oxidizing or reducing conditions. The presence of various reduced chemical species in soil is often a better indicator of the redox status of a soil than measured E_h values.

The soil pH controls many of the important variables in the soil environment such as the size and nature of the microbial community, the composition of the soil solution, the composition of the soil organic matter, and the activity of enzymes. The measurement of soil pH is one of the most common and important analytical techniques in soil biochemistry. It has often been observed that the optimum pH of an enzyme is 0.5-2.0 units higher when assayed in soil as compared to that in solution.

Temperature plays an important role in controlling the growth and activity of microorganisms, partial pressure of gases, redox potentials, and diffusion of solutes and gases. Biological activity can occur in soils at greater temperature extremes than would be expected in studying isolated systems due to the heterogeneity and complexity of soils.

The microbial biomass comprises approximately 1 to 4 percent of the total organic matter in soil. The microbial cell in soils can be viewed as being in a dynamic state, adapting to shifts in the environment. The majority of microorganisms cannot survive the transfer from the soil onto a culture medium making it difficult to assign a specific function to a specific organism in the soil. The soil biochemist, therefore, is not so much interested in an accurate description of each of the species of soil microorganisms but in their biochemical activity.

Soil chemical, physical and biological properties can vary considerably between soil samples. This happens not only between soils from different mapping units but also between soils within a small unit area, or in individual soil horizons. What is generally measured in the soil is the sum of reactions at individual locations or micro-sites in the soil. However, to understand the soil biochemistry, one must have a good insight of the activities that can differ from one micro-site to another.

The cycling of nutrients within nature is necessary to avoid that large pools of compounds are formed containing one or more basic nutrients. The cycling of carbon is a good example of such a situation. The energy tied up in the carbon-hydrogen bond is constantly being released by microorganisms so that they can harness this energy for their own growth and reproduction. The cycling of nitrogen, phosphorus, sulfur, and several metals is often based on the transformations of carbon. Several factors affect the cycling of nutrients in soil. Parent material composition, climate, and activities of man (cultivation, fertilizer additions, pollution, etc.) all affect the system, causing changes in the rate of cycling and in pool sizes of individual compounds of each nutrient.

The steady increase of population and human activities causes a continuous accumulation of organic and inorganic pollutants such as xenobiotics and heavy metals in the soil and, therefore, much more attention is nowadays paid to soil remediation. Many biochemical processes in soil can change the availability and toxicity of such contaminants, and many enzymes including hydrolases and oxidoreductases are responsible for the degradation of organic contaminants and the formation of humic polymers. Some organic ligands may bind heavy metals and reduce their availability to plants and microorganisms. A number of heavy metals could be transformed to gaseous forms via enzymatic reactions. It is certain that soil biochemistry plays more and more important roles in the bioremediation and restoration of soils.

Chemistry of Soil Organic Matter

Pools of Organic Matter in Soils

Soil organic matter includes litter, light organic fraction, microbial biomass, water-soluble organics and humus. Litter is defined as the macro-organic matter on the soil surface. This pool is particularly important to the cycling of nutrients in forest soils and natural grasslands. In cultivated soils, crop residues are incorporated into the soil well in advance of planting, and the litter becomes part of the light fraction. The light fraction consists mainly of plant residues in varying stages of decomposition; it is used as an indicator of the changes in labile organic matter as affected by tillage, crop rotation and environmental conditions affecting microbial activity. The size of the light fraction is considerably less than that of stable humus, but can constitute as much as 30 percent of the organic matter in some soils.

The microbial biomass plays a dual role in the soil as an agent for the decomposition of plant residues and as a labile pool of nutrients. Water-soluble organics are important both from the view of plant nutrition and the environment (binding of metals and xenobiotics). In most agricultural soils, the bulk of the organic matter occurs as stable humus.

Formation of Humic Substances

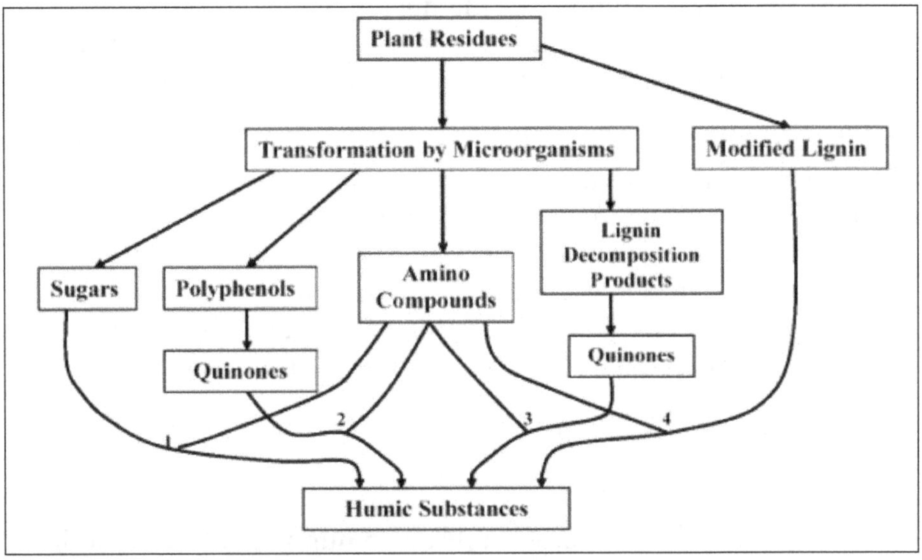

Pathways for the formation of humic substances.

Humic substances are a series of relatively high-molecular-weight, yellow to black colored substances formed by secondary synthesis reactions. The formation of humic substances is one of the least understood aspects of soil biochemistry. Figure above depicts several pathways for the formation of humic substances during the decay of plant and animal remains in soil. The classical theory is that humic substances represent modified lignin (pathway 4). However, the majority of present-day investigators favor a mechanism involving quinones (pathways 2 and 3). In practice, all four pathways must be considered as likely mechanisms for the synthesis of humic and fulvic acids in nature, including sugar-amine condensation (pathway 1). All pathways may be involved in humus formation, but not to the same extent in all soils or in the same order of importance. However, it is difficult to say which pathway is more important in a specific soil.

The Lignin Theory

For many years it was thought that humic substances were derived from lignin, though there is still a lot of uncertainty about this issue. According to this theory, lignin is incompletely utilized by microorganisms and the residuum becomes part of the soil humus. Modification in lignin includes loss of methoxyl (OCH_3) groups with the generation of o–hydroxyphenols and oxidation of aliphatic side chains to form COOH groups. The nitrogen contained in humic acids results from the condensation of modified lignin with protein, the latter being a product of microbial synthesis. Stabilization of the protein is believed to occur through formation of a Schiff base.

$$(\text{Modified lignin})-\text{CHO} + \text{RNH}_2 \rightarrow (\text{Modified lignin})-\text{CH}=\text{NR}+\text{H}_2\text{O}$$

In aerobic soils, lignin may be broken down into low-molecular-weight products prior to humus synthesis. Oxygen is required for the microbial depolymerization of lignin, and fungi that degrade lignin are not normally found in excessively wet sediments. Therefore, modified lignins will make a major contribution to the humus of poorly drained soils.

Demethylation and oxidation of side chains lead to the formation of products enriched in acidic functional groups (COOH and phenolic OH) with lower C but higher oxygen contents than the original lignin. The COOH groups arising from the side chains of lignin are not sufficient to account for the high COOH content of humic and fulvic acids. Additional COOH groups may arise from ring cleavage of aromatic components of lignin. The o–dihydroxy-benzene units resulting from demethylation of lignin would be subject to oxidation to quinones capable of undergoing condensation reactions with NH_3 and amino acids.

The Polyphenol Theory

In pathway 3, lignin still plays an important role in humus synthesis, but in a different way. Phenolic aldehydes and acids released from lignin during microbiological attack undergo an enzymatic conversion to quinones, which polymerize in the presence or absence of amino compounds to form humic-like macromolecules. Pathway 2 is somewhat similar to pathway 3 except that the polyphenols are synthesized by microorganisms from non-lignin C sources. The polyphenols are then enzymatically oxidized to quinone and converted into humic substances.

Quinones of lignin origin, together with those synthesized by microorganisms, are the major building blocks from which humic substances are formed. In this model, the first step consists of the breakdown of all plant biopolymers into their monomeric structural units, some of which polymerize enzymatically to produce humic molecules of increasing complexity. The formation of humic substances would thereby follow the order:

$$\text{Fulvic acid} \rightarrow \text{humic acid} \rightarrow \text{components of humin}$$

Possible sources of phenols for humus synthesis include lignin, microorganisms, uncombined phenols in plants, glycosides and tannins. Polyphenols are not stable and subject to further decomposition by bacteria, actinomycetes, and fungi. They may also undergo recombination, either alone or with other organic molecules, after conversion of quinones. The conversion of phenols to quinones is likely to be carried out by polyphenoloxidase enzymes. For the products derived from lignin, oxidation to quinones and consequent polymerization to humic substances must be preceded by demethylation to account for the low OCH_3 but high phenolic OH contents of humic and fulvic acids.

Other changes include oxidation of aldehyde components, decarboxylation and hydroxylation, and coupling of intermediates. Self-condensation of quinones or free radicals are greatly enhanced in the presence of amino compounds, such as amino acid, peptides, and proteins.

Sugar-amine Condensation

For pathway 1, humus is supposed to be formed from sugars. Reducing sugars and amino acids, formed as by-products of microbial metabolisms, undergo a non-enzymatic polymerization to form brown nitrogenous polymers. Unlike the lignin theory, the starting material consists of low molecular weight organic compounds, from which large molecules are formed through condensation and polymerization. Sugars and amino acids are produced in abundance in the soil through microorganism activity. This theory could explain the formation of humic substances in environments where lignin and lignin degradation products are not prevalent.

Many different types of reactions can lead to the production of dark colored pigments in soil. The major pathway for the formation of humic substances in most soils appears to be through condensation reactions involving polyphenols and quinines. Polyphenols derived from lignin, or synthesized by microorganisms, are enzymatically converted to quinones, which undergo self-condensation or combine with amino compounds to form N-containing polymers. The number of precursor molecules is large and the number of ways in which they combine is astronomical, thereby accounting for the heterogeneous nature of the humic material in any given soil.

Soil Enzymes

The transformations of many chemical compounds in soil are catalyzed by proteins called enzymes. Like all catalysts, enzymes increase the rate of a chemical reaction without themselves undergoing permanent alteration. Soil enzymes are composed of four classes: hydrolases, lyases, oxidoreductases, and transferases. Hydrolases catalyze the hydrolysis of a variety of bonds, such as esters, glycosides and peptides. Lyase enzymes catalyze the non-hydrolytic removal of a group from a substrate with the resulting formation of a double bond. Enzymes involving the catalysis of oxidationreduction reactions are called oxidoreductases. Enzymes that transfer a specific grouping from one molecule to another are transferases. Some soil enzymes catalyze reactions both inside and outside the organisms; some can only function outside the producing cell because of the large size of their target substrate; and others, being involved in the central aspects of metabolism, do not function extra-cellularly. Only a limited number of enzymes have been extracted from soil.

Classification by Location

The overall enzyme activity of a soil is a composite of various intra- and extra-cellular enzyme components produced by microorganisms or derived from animal and plant sources. Intra-cellular enzymes may exist in different cellular compartments of living cells, such as the cytoplasm or periplasm, or they may be associated with the cell membrane and wall. Extra-cellular enzymes are produced and secreted by living cells and operate at a distance from the parent cell, either as free enzymes in the liquid phase or as enzymes still associated with the external surface of root epidermal or microbial cell wall.

Classification by Function

Two groups of enzyme activities appear to predominate in the soil: hydrolases and oxido-reductases. Hydrolases, such as polysaccharidases and proteinases, transform macromolecular substrates into smaller compounds. The depolymerization of carbohydrates and proteins and the subsequent mineralization of their monomeric products have special importance in the biological cycling of both carbon and nitrogen in soils. Polysaccharidases (e.g. amylases, cellulases, xylanases, dextranases) catalyze the first stage of the depolymerization of polysaccharides to oligo- and monosaccharides. These, in turn, can be assimilated by living organisms or undergo further decomposition in soil by the action of hydrolytic enzymes such as α- or β-glucosidases and α- or β-galactosidases. The final hydrolysis products of these enzymes, such as glucose or galactose, constitute important energy sources for soil microorganisms. Hydrolysis of saccharose by invertase is another significant reaction of carbohydrate metabolism in soils. The presence of invertase has been correlated with microbial activity and is regarded as a useful "fertility index".

Proteins in soils are hydrolyzed to peptides and amino acids by proteinase and peptidase activities. Urease catalyzes the hydrolysis of urea into ammonia and carbon dioxide, which is an important reaction in the soil nitrogen cycle. In the P and S cycles the inorganic and organic forms are closely associated through mineralization and immobilization processes mediated by abiotic and biotic activities.

Herbicide- and insecticide-hydrolyzing activities, such as those of propanilacylamidases or malathion- and parathion-esterases, have been studied in various soils. These enzymes which, in the presence of active co-factors, catalyze the oxidation of numerous organic compounds by removal of electrons and hydrogen, are believed to be localized only within intact viable cells. Treatment of soils with bacteriostatic or bactericidal compounds destroys their activity.

The transformation of atmospheric H_2 in moist soils is assumed to be catalyzed by hydrogenases. Reduced organic compounds, mostly phenolic, may be oxidized by the action of peroxidase. The presence of hydrogen peroxide, which is transformed to water and O_2, is required during this reaction. Hydrogen peroxide, a potential toxic intermediate of respiration, may also be destroyed by catalase, which produces molecular oxygen and water. This decomposition of hydrogen peroxide makes catalase an important enzyme of soil biochemical activity.

Peroxidases like manganese and lignine peroxidases and polyphenol oxidases such as laccases and tyrosinases catalyze the oxidation of a variety of phenols, aromatic amines, and other compounds. The oxidative activities of these enzymes, such as the oxidative coupling reaction, play an important role in humus formation. Oxidoreductive enzymes also participate in the degradation or incorporation of xenobiotic substances into soil organic matter.

Soil Organic Matter

Soil organic matter (SOM) is the organic matter component of soil, consisting of plant and animal detritus at various stages of decomposition, cells and tissues of soil microbes, and substances that soil microbes synthesize. SOM provides numerous benefits to the physical and chemical properties of soil and its capacity to provide regulatory ecosystem services. SOM is especially critical for soil functions and quality.

The benefits of SOM result from a number of complex, interactive, edaphic factors; a non-exhaustive list of these benefits to soil function includes improvement of soil structure, aggregation, water retention, soil biodiversity, absorption and retention of pollutants, buffering capacity, and the cycling and storage of plant nutrients. SOM increases soil fertility by providing cation exchange sites and being a reserve of plant nutrients, especially nitrogen (N), phosphorus (P), and sulfur (S), along with micronutrients, which the mineralization of SOM slowly releases. As such, the amount of SOM and soil fertility are significantly correlated.

SOM also acts as a major sink and source of soil carbon (C). Although the C content of SOM varies considerably, SOM is ordinarily estimated to contain 58% C, and "soil organic carbon" (SOC) is often used as a synonym for SOM, with measured SOC content often serving as a proxy for SOM. Soil represents one of the largest C sinks on Earth and is significant in the global carbon cycle. Therefore, SOM/SOC dynamics and the capacity of soils to provide the ecosystem service of carbon sequestration through SOM management have received considerable attention recently.

The concentration of SOM in soils generally ranges from 1% to 6% of the total mass of topsoil for most upland soils. Soils whose upper horizons consist of less than 1% of organic matter are mostly limited to deserts, while the SOM content of soils in low lying, wet areas can be as great as 90%. Soils containing 12% to 18% SOC are generally classified as organic soils.

It can be divided into 3 genera: The living biomass of microbes, fresh and partially decomposed detritus, and humus. Surface plant litter, i. e., fresh vegetal detritus, is generally excluded from SOM.

The primary source of SOM is vegetal detritus. In forests and prairies, for example, different organisms decompose the fresh detritus into simpler compounds. This involves several stages, the first being mostly mechanical, and becoming more chemical as decomposition progresses. The microbial decomposers are included in the SOM, and form a food web of organisms that prey upon each other and subsequently become prey.

There are also other herbivores that consume fresh vegetal matter, the residue of which then passes to the soil. The products of the metabolisms of these organisms are the secondary sources of SOM, which also includes their corpses. Some animals, like earthworms, ants, and centipedes contribute to both vertical and horizontal translocation of organic matter. Additional sources of SOM include plant root exudates and charcoal.

Composition of Vegetal Detritus

The water content of most vegetal detritus is in the range of 60% to 90%. The dry matter consists of complex organic matter that is composed primarily of carbon, oxygen, and hydrogen. Although these three elements make up about 92% of the dry weight of the organic matter in soil, other elements are very important for the nutrition of plants, including nitrogen, phosphorus, potassium, sulfur, calcium, magnesium, and many micronutrients.

Organic compounds in vegetal detritus include:

- Carbohydrates that are composed of carbon, hydrogen, and oxygen, and range in complexity from rather simple sugars to the large molecules of cellulose.
- Fats that are composed of glycerids of fatty acids, like butyric, stearic, and oleic. They also include carbon, oxygen, and hydrogen.
- Lignins that are complex compounds, form the older parts of wood, and also are composed primarily of carbon, oxygen, and hydrogen. They are resistant to decomposition.
- Proteins that include nitrogen in addition to carbon, hydrogen, and oxygen; and small amounts of sulfur, iron, and phosphorus.
- Charcoal, which is elemental carbon that is derived from incomplete combustion of organic matter. It is resistant to decomposition.

Decomposition

Vegetal detritus in general is not soluble in water and therefore is inaccessible to plants. It constitutes, nevertheless, the raw matter from which plant nutrients derive. Soil microbes decompose it through enzymatic biochemical processes, obtain the necessary energy from the same matter, and produce the mineral compounds that plant roots are apt to absorb. The decomposition of organic compounds specifically into mineral, i. e., inorganic, compounds is denominated "mineralization". A portion of organic matter is not mineralized and instead decomposed into stable organic matter that is denominated "humus".

The decomposition of organic compounds occurs at very different rates, depending on the nature of the compound. The ranking, from fast to slow rates, is:

- Sugars, starches, and simple proteins.
- Proteins.
- Hemicelluloses.

- Cellulose.
- Lignins and fats.

The reactions that occur can be included in one of 3 genera:

- Enzymatic oxidation that produces carbon dioxide, water, and heat. It affects the majority of the matter.
- A series of specific reactions liberates and mineralizes the essential elements nitrogen, phosphorus, and sulfur.
- Compounds that are resistant to microbial action are formed by modification of the original compounds or by microbial synthesis of new ones to produce humus.

The mineral products are:

Element	Mineral Products
Carbon	CO_2, CO_3^{2-}, HCO_3^-, CH_4, C
Nitrogen	NH_4^+, NO_2^-, NO_3^-, N_2 (gas), N_2O (gas)
Sulfur	S, H_2S, SO_3^{2-}, SO_4^{2-}, CS_2
Phosphorus	$H_2PO_4^-$, HPO_4^{2-}
Others	H_2O, O_2, H_2, H^+, OH^-, K^+, Ca^{2+}, Mg^{2+}, etc.

Humus

As vegetal detritus decomposes, some microbially resistant compounds are formed, including modified lignins, oils, fats, and waxes. Secondly, some new compounds are synthesized, like polysaccharides and polyuronids. These compounds are the basis of humus. New reactions occur between these compounds and some proteins and other products that contain nitrogen, thus incorporating nitrogen and avoiding its mineralization. Other nutrients are also protected in this way from mineralization.

Humic Acids/Substances

Humic acids/substances are classified into 3 genera based on their solubility in acids and alkalis, and also according to their stability:

- Fulvic acid is the genus that contains the matter that has the lowest molecular weight, is soluble in acids and alkalis, and is susceptible to microbial action.
- Humic acid is the genus that contains the intermediate matter that has medial molecular weight, is soluble in alkalis and insoluble in acids, and has some resistance to microbial action.
- Humin is the genus that contains the matter that has the greatest molecular weight, is the darkest in color, is insoluble in acids and alkalis, and has the greatest resistance to microbial action.

Function in Carbon Cycling

Soil has a crucial function in the global carbon cycle, with the global soil carbon pool estimated to be 2,500 gigatons. This is 3.3 times the amount of the atmospheric pool at 750 gigatons and 4.5 times the biotic pool at 560 gigatons. The pool of organic carbon, which occurs primarily in the form of SOM, accounts for approximately 1,550 gigatons of the total global carbon pool, with soil inorganic carbon (SIC) accounting for the remainder. The pool of organic carbon exists in dynamic equilibrium between gains and losses; soil may therefore serve as either a sink or source of carbon, through sequestration or greenhouse gas emissions, respectively, depending on exogenous factors.

Chemistry of Soil Organic Matter

In the classical definition the term “humus” refers to the components of SOM that have been transformed to the extent that they bear no morphological resemblance to the materials of origin. Hayes and Swift subdivided these ‘humified’ materials into: (1) Amorphous, polymeric, brown-colored humic substances which are differentiated on the basis of solubility properties into humic acids (HAs), fulvic acids (FAs), and humins (HUs); and (2) compounds belonging to recognizable classes, such as polysaccharides, polypeptides, altered lignins, etc. These can be synthesized by microorganisms or arise from modifications of similar compounds in the original debris.

The more-widely used definitions consider humic substances (HSs) to be “a series of relatively high molecular weight, yellow to black colored substances formed by secondary synthesis reactions”, and “a category of naturally occurring, biogenic, heterogeneous organic substances that can generally be characterized as being yellow to black in colour, of high molecular weight, and refractory”. These definitions are vague, and some of the statements are being questioned at the present time. Nonetheless, the subdivisions outlined are useful for operational purposes.

Humic acids (HAs) are the components of HSs which are soluble in basic media but precipitated in acidic media at pH 1 (water chemists use pH 2 as the cutoff, at which point effectively all of the perceptible components from soils are precipitated). Many non-humic materials (such as some proteins, for example) are precipitated under similar conditions. In the classical definitions fulvic acids (FAs) are the components of alkaline extracts that remain in solution after a medium is acidified. Clearly that fraction would also contain many of the components of humus that belong to recognizable classes. Hence the term “fulvic acid fraction” is used for materials that remain in solution in the acidified medium. FA fractions from which the standard and reference FAs of the International Humic Substances Society (IHSS) were obtained were passed onto XAD-8 ((poly) methylmethacrylate) resin. Polar non-humic components (such as saccharides) are considered to pass through the resin, while FAs are retained by the resin.

The materials that pass through the resin may be retained by XAD-4 (styrene-divinylbenzene); these are defined as XAD-4 acids.

Humins (HU) is the component of HSs that is not soluble in aqueous media at any pH value. It is a major component of the soil humic fraction and represents more than 50% of the organic C (OC) of soil and more than 70% of the OC in lithified sediments. It is often considered to be the component of HSs associated with mineral colloids that is not extracted in an aqueous base. However, lignin residues that have not been sufficiently oxidized to provide hydratable functionalities in basic media, as well as algenans, cutans, and suberans, can contribute to the humin fraction. On the basis of considerations of solubility in aqueous media, any HSs material that is extracted in an organic solvent, after an exhaustive extraction, in an aqueous base can be a considered to be humin.

Aspects of Genesis

Stevenson outlined 3 pathways for the formation of HSs. In 1, amino compounds from microbial synthesis react with modified lignins. In the second, polyphenols oxidized to quinones react with amino compounds, and in the third, reducing sugars react with amino compounds to synthesize complex dark-colored macromolecules. Two general synthesis processes are involved. One, the degradation process, involves modifications of relatively intractable plant components such as lignins, suberins, cutins, etc., and also includes microbial products such as paraffinic macromolecules in algae and plant cuticles, such as the enzymatic depolymerization by activated oxygen free radicals (e.g., superoxide radicals) of species such as lignins and tannins. Stevenson also indicated how microorganisms secrete oxidative enzymes that depolymerize plant polymers.

Oxidative transformations give rise to a variety of reactive functionalities, especially carbonyl/carboxyl groups. In the case of lignin, for example, the unaltered parts are relatively non-polar (hydrophobic), while the oxidized segments are hydrophilic. Thus, transformed lignin is an amphiphile (with separate hydrophobic and hydrophilic parts) and can exhibit surface-active properties. Zech et al. discussed modifications to plant residues during the humification process, and these, among other changes, include oxidation and transformation of lignin side chains, the demethylation of lignins, the removal of carbohydrate materials, and the conversion of catechol-like products to quinones.

The second, or abiotic condensation (AC) process, involves the complete breakdown of bonds linking the component molecules of biopolymers and the re-synthesis of macromolecular products through interactions between reactive molecules. The third pathway of Stevenson is also an abiotic process. Reducing sugars released from plant and animal sources react with amino compounds (amino acids and peptides) also released in degradation processes. This process, known as the Maillard reaction, gives rise to macromolecular melanoidins.

Isolation and Fractionation of Humic Substance

Isolation of Humic Substances from Soil

Traditionally, aqueous solvents are used for isolating HSs from soils. Some organic solvents can be applied, Hayes concluded that "good" organic solvents have electrostatic factor (or the product of the relative permittivity and dipole moment) values exceeding 140 and pK_{HB} (a measure of the strength of the solvent as an acceptor in hydrogen bonding) values of > 2. Dimethyl formamide (DMF) and dimethyl sulfoxide (DMSO) meet these requirements, but there are problems with the recoveries of conventional fractions of HSs from these solvents. Hayes also discussed solubility parameter data to explain differences in the abilities of organic solvents to solvate HSs. The best of these solvents for H^+ -HAs have d_p (dispersion force), dh (hydrogen bonding), and d_b (proton acceptor) values exceeding 6, 5, and 5, respectively. Solvation is greatest when the product d_a (solvent) × d_b (solute), or vice versa, is greatest. Water has large values for d_h, d_a, and d_b, which reflect the extents of its self-association because of hydrogen bonding. It follows that the d_a and d_b values of H^+ -HAs are not sufficient to disrupt the attractive forces. Although water satisfies all of these criteria, it is a poor solvent for H^+ - and divalent- and polyvalent cation-exchanged HAs.

Standard isolation processes use water as the solvent. These involve extraction in basic and neutral salt solutions using, for example, 0.1 M NaOH after the sample is H^+ - ion exchanged, usually with HCl (up to 1 M), or when not H+ -exchanged, 0.1 M NaOH + 0.1 M sodium pyrophosphate ($Na_4P_2O_7$, Pyro) is used. The acid treatment displaces chargeneutralizing cations, and the acidic functionalities in the H^+ -exchange system dissociate when the pH is raised above the pK_a of the functionalities. Then the conjugate bases solvate in water. Neutralization with H_3PO_4, which serves as a neutral salt capable of complexing divalent and polyvalent metals, gives rise to inter- or intramolecular bridging effects that render the HSs insoluble in water. These metals are exchanged for Na^+, and the Na^+ -carboxylates are solvated. However, HSs with low carboxyl charge densities will not dissolve, but will do so at higher pH values should sufficient phenolic functionalities be contained in the humic molecules. Na^+ -Pyro, at 0.1 M, has a pH value of 10.6, and the pK_a values of the phenolic functionalities are in the range of 8 to 10. Weaker acidic functionalities, such as enols, contribute to the isolates at pH 12.6.

Hayes referred to 0.1 M NaOH + 6 M urea for exhaustive extractions following extraction with 0.1 M NaOH. The compositions of the isolates were broadly similar to those isolated at pH 12.6, indicating that the urea may have liberated humic materials trapped within or hydrogen-bonded to the humin matrix. Hayes et al. carried out exhaustive and sequential extractions using distilled water, Pyro. at pH 7, Pyro. at pH 10.6, and Pyro. + 0.1 M NaOH.

Recently, Hayes referred to the use of DMSO + 6% concentrated H_2SO_4 following extraction with NaOH + urea. The extract in that instance significantly differed from prior extracts The procedure used to isolate the standards of the IHSS employs a 0.1 M

HCl/0.3 M HF treatment to remove finely divided inorganic soil colloids, and thereby provide materials with acceptable ash contents. HCl/HF treatment is repeated as often as necessary to lower the ash content to < 1%. This treatment, followed by dialysis, leads to considerable losses of the humic fractions. Clapp and Hayes and Hayes et al. found that filtration through partially clogged 0.45- or 0.2- μm filters effectively lowered the ash contents of the fractions, and of course this procedure avoids degradation of HSs.

Fractionation of HSs in the Soil

Uses are made of polarity, solubility, molecular size, and charge density differences for the fractionation of HSs. In principle, humic fractions that have weakly dissociable acids should be the first to precipitate as the pH is lowered, and the strongest acids would be last. Uses can be made of salts to give 'salting in' and 'salting out' effects, and of poor solvents such as ethanol and methanol to fractionally precipitate HAs from alkaline solutions.

A procedure, based on that of Malcolm and MacCarthy, sufficiently dilutes SOM extracts (50~100 ppm) so that the DOM stays in solution when the pH is adjusted to 2 (dilution to < 50 ppm is preferable). This solution is passed onto XAD-8 ((poly) methylmethacrylate) and XAD-4 (styrenedivinylbenzene) resins in tandem (packed in 0.01 M acid). After back-elution with dilute NaOH, the HAs are precipitated at pH 1 and separated from the FAs. Hayes et al. modified this approach and exhaustively extracted with 1 solvent before proceeding to the next in the series: Water; 0.1 M Pyro at pH 7.0; 0.1 M Pyro at pH 10.6; and 0.1 M NaOH + 0.1 M Pyro at pH 12.6. In this approach, the combined extracts (for each solvent system) are diluted to < 30 ppm, pressure-filtered through 0.2 μm pore size filters, the pH adjusted to 2, and the solutions passed through XAD-8 and XAD-4 resins in tandem. The more-hydrophobic components, retained by the XAD-8 resin, are back-eluted in 0.1 M NaOH, and the pH of the centre cut is adjusted to 1 (using 6 M HCl).

Precipitation of the HAs is allowed to take place overnight at 4 °C. After centrifugation, the FAs are desalted by passing them onto an XAD-8 resin, then passing distilled water through it until the conductivity is < 100 S cm^{-1}, back-eluting with 0.1 M NaOH, passing the eluate slowly through an IR-120 (H^+-exchanged, sulphonated polystyrene) resin column, and finally freeze-drying. To desalt the HAs, the solution is dialyzed against distilled water until the dialysate is chloride free, and then it is freeze-dried. Alternatively, the HAs may be dissolved in a base and diluted until the concentration is < 30 ppm. The pH of the solution is adjusted to 2, it is passed slowly onto XAD-8 (in 0.01 M HCl) resin and treated as for the FAs. Hayes et al. observed that in some instances, precipitates were formed in the pH 2.5 to 2.0 range, and the compositions of these precipitates significantly differed from the HAs isolated from the XAD-8 resin. XAD-4 acids cannot be considered to be valid components of HSs. These would be components of the HAs and FAs in the classical definitions, but should be regarded as non-humic materials in associations with the true humics in the operational definitions. We have observed that hydrocarbon-, fatty acid-, and wax-type materials are sorbed onto the cellulose acetate membrane during the filtration process, and these would also normally be included with the humic (HAs) fractions in the classical operational definition.

Composition and Structures

In the humic sciences, composition refers to the functionalities and to the component molecules. In the classical chemical sense, structures of macromolecules include the size and shape, the sequences in which component molecules are linked, and the nature of the linkages, i.e., the associations between molecules that comprise their structures.

Analytical data are most meaningful when the samples analysed and compared have been subjected to the same isolation and fractionation procedures. A set isolation and fractionation procedure was followed during the isolation of IHSS standards. However, additional fractionations based on the charge density and size differences give subfractions with different compositions. Swift updated the wet chemical and some spectroscopic procedures for determining of functional groups of HSs, and procedures are given in IHSS publications.

Although HSs are heterogeneous mixtures, the evidence suggests that substances from the same soil types and formed under the same environmental conditions are broadly similar. Nevertheless, it is unlikely that any 2 true humic molecules even in the same batch are exactly the same should the molecules have molecular weight (MW) values of even 3000 Da. However, distinct compositional differences can be expected between different fractions (HAs/FAs/humins) isolated from the same humic source, and strong differences may be observed between the humic fractions from different sources, soil types, and climates. Some general reactivities, regardless of source, will likely be broadly similar.

There is disagreement about the extents to which HSs can be considered to be macromolecular. Macro-molecularity can, to some extent at least, be determined by the degree to which there are, if any, covalent links between HS molecules and peptides and oligo and polysaccharides. There should be close associations between these biological molecules and HSs (and there are sufficient interactive functionalities in all of these molecules to allow such interactions to take place), then the sugars and peptides can be considered to be akin to prostethic groups of proteins.

Sizes and Shapes of Humic Molecules

The classical concepts regard HSs as macromolecles with molecular weight values up to 1.5×10^6 Da. However, a second concept suggests that HSs are only pseudomacromolecles, and the macromolecular-like properties arise from associations of smaller molecular species in micellar or pseudomicellar structures, or through other forms of molecular associations possibly involving non-polar molecules such as long-chain hydrocarbons, fatty acids/esters/suberin-type, lipoprotein, components.

The different arguments with regard to sizes and shapes of humic molecules discussed here. Considerations are given to the concept of HSs as micellar structures, and considerations of the micellar-type associations of humic molecules have considerable support. Micelle formation requires HSs to have amphiphilic properties wherein molecules have hydrophobic (nonpolar) and hydrophilic (polar) parts. At low concentrations, amphiphilic molecules exist solely as single unit species, but at concentrations above the so-called critical micelle concentration (CMC), or the concentration above which a surfactant molecule in solution will form an ordered aggregate or micelle, these will form compact, ordered aggregates or micelles. In micelles the hydrophobic groups are orientated towards each other and the interior, and the hydrophilic functionalities are on the outside. In special cases, polar groups can be found on the inside, and in such circumstances, interactions between these, such as hydrogen bonding, will decrease the hydrophilic character of the polar groups. Also the hydrophilic groups (inside or outside) in the micellar structures can interact through hydrogen bonding and dipole-dipole interactions with the solvent, water. Entropy increases as the water structure is disrupted and decreases with micelle formation. There is, however, a favorable increase in energy when roughly spherical micelles are formed (with the radius approximately equivalent to the length of the hydrophobic group in linear surfactants) at solution concentrations < 10 times the CMC. Non-spherical micelles can form at higher concentrations. The free energy of the system increases when the structure of the solvent is disrupted during the addition of the surface active materials. This increase is minimized when the molecules are concentrated at the surface and are oriented so that their hydrophobic groups are directed away from the liquid.

The micelle and hydrogen bonding concepts are, of course, forms of molecular associations. The micelle concept also includes hydrophobic bonding through associations of the non-polar functionalities.

Compositions of Humic Substances

Information from Chemical Degradation Procedures

Arguably, chemical degradation procedures and nuclear magnetic resonance (NMR) provide the most useful information about the molecules that compose HSs. The components of biological polymers have labile linkages that are readily hydrolysed. Although the information we have indicates that the molecular components or 'building blocks' of HSs are relatively simple, these molecules or units are linked by bonds that are, for the most part, difficult to cleave. For example, the "backbone" structures, or the major compositional units are linked by carbon-carbon bonds, by either linkages, or by other types of linkages that are degraded only when high inputs of energy are used. Up to 50% of the masses of soil HAs can be lost as CO_2, especially from activated carboxyl groups, such as ketone acids, from hydroxybenzenecarboxylic acids, and as soluble molecules (e.g., sugars, amino acids, small amounts of purine and pyrimidine bases, and phenolic substances). These estimates were made for HAs fractions, which of course have saccharide and peptide materials especially associated with HAs. The total acidities of soil HAs are not decreased by hydrolysis because other acid groups are formed (e.g., from esters and lactones).

Hayes and Swift and Clapp et al. compiled a list of more than 100 products that have been identified in the digests of a variety of oxidative chemical degradations of HSs. Many additional products have been identified in the digests of reductive degradation procedures and as pyrolysis products in pyrolysis/gas chromatography/mass spectrometry (Pyr/GC/MS) degradation studies. At one time it was considered that the digest products were "building blocks" of humic macromolecules, but Hayes and Swift took into account the mechanisms involved in the degradation by the different chemical reagents and showed that 1 parent compound in the macromolecule can give rise to several digested products. The products depend on the type and concentration of reagent, the temperature, the reaction time, and so on. Most of the compounds identified in oxidative degradations are acids, as would be expected on the basis of mechanistic considerations. The aromatic acids range from benzenedi- to benzehexacarboxylic acids, and there is abundant evidence for monoto tri-hydroxy (or methoxy), and hydrocarbon substituents on the aromatic rings.

Fused aromatic structures are major digest products in zinc dust distillation and fusion reactions, and some of these have heterocyclic N digest products as listed by Hayes and Swift, Stevenson, and Clapp et al. Concepts of the structures of HSs were influenced for a time after such structures were identified in digests.

The abundance of benzenecarboxylic acids in the digests of alkaline permanganate oxidative degradations can, of course, arise from oxidations of fused aromatic components. It is now agreed that these may be artefacts. In the context of oxidative degradation, alkaline cupric oxide does not degrade fused aromatic structures to benzene polycarboxylic acids, and because these acids are contained in considerable abundances in the alkaline cupric oxide digests of HSs, it is not likely that fused aromatic structures can be considered to contribute to any significant extent to humic structures. However, these could arise from partially associated char materials that behave like HSs. Carbonylation reactions can produce benzenecarboxylic acids, and this possibility has not been resolved for the alkaline degradations of HSs.

Aliphatic dicarboxylic acids, ranging from ethanedioic to decanedioic, are abundant in oxidative digests, and there is also evidence for aliphatic tricarboxylic acids. Carboxyls could arise from cleavages of unsaturated groups in aliphatic chains and could also form from oxidizable functionalities, such as alcohols and aldehydes. Inevitably some are indigenous to the aliphatic structures.

Sodium sulphide at elevated temperatures is used for the delignification of wood, Hayes and Swift devised a procedure for degradation of HSs using a saturated solution (10%) of Na_2S at 250 °C in an autoclave. The aromatic compounds identified variously had 1 or 2 methoxy substituents (hydroxyl would have been converted to methoxyl in the methylation process needed for analysis by GC-MS), 1 or 2 methyls or other aliphatic substituents, and with the exception of phthalic acid, there is never more than 1 carboxyl group attached to the benzene nucleus. This contrasts with benzenepolycarboxylic acids found in digests in which permanganate and alkaline cupric oxide are used.

Information from pyrolysis

There are increasing uses of pyrolysis (Pyr) procedures for studying the component molecules of HSs. Because of the large energy inputs in Pyr processes, most of the components of pyrolyzed macromolecules are drastically altered. Hence, awareness of the types of structures which can give rise to pyrolyzate products is essential for making plausible predictions of their structures. Pyrolyzates of soil HSs yield compounds indicative of origins in carbohydrates, polypeptides, lignins, polyphenols, and fatty acids and esters products, with likely origins in lignins and in microbially synthesized polyphenol substances.

Information from Spectroscopic Procedures

Nuclear magnetic resonance (NMR) is now a powerful spectroscopic technique which provides information about functionalities. Significant progress has been made since techniques such as multiple-pulse Fourier-transformation NMR, dipolar decoupling, cross-polarization nuclear induction spectroscopy, and cross-polarization magic angle spinning (CPMAS), direct polarization (DP) NMR, and dipolar dephasing (DD), and chemical shift anisotropic (CSA) techniques were introduced. These techniques enable better assignments of functional groups and allow some degree of quantification to be achieved.

Solid-state NMR has greatly contributed to our understanding of the compositions of HSs. Resonance bands in the 10-45 (alkyl C), 45-65 (O-alkyl), 65-110 (ethers, hydroxyls, sugars), 110-140 (aromatic C), 140-160 (phenols, O-aryl), 160-190 (carboxyl, esters, amides), and 190-220 ppm (carbonyl of aldehydes, ketones) bands allow predictions to be made about the functionalities of humic components. Application of DD spectroscopy allows predictions of the extents of aromatic substitutions, and highlights methyl functionalities (including O-methyl). The CSA process suppresses aromatic functionalities and clearly distinguishes the aromatic C (with resonance at 105 ppm).

The development of high field magnets (of 500, 600, and 800 MHz) and multidimensional experiments has enabled liquidstate NMR to solve the problems associated with overlap in 1-dimensional (1D) spectra. Proton NMR (^1H NMR) in combination with ^{13}C NMR in 2D ^1H-^{13}C heteronuclear multiple quantum coherence (HMQC) and heteronuclear multiple bond coherence (HMBC) correlation techniques; correlation spectroscopy (COSY); 2D homonuclear hartmann hahn (HOHAHA); total correlation spectroscopy; and rotating frame overhauser enhancement spectroscopy (ROESY) or nuclear overhauser effect spectroscopy (NOESY), and diffusion ordered spectroscopy (DOSY) have allowed considerable detail to be obtained about the functionalities, components, and associations of molecules in HSs.

Applications of DOSY showed that operationally defined HAs are mixtures of proteinaceous-, carbohydrate-, and lignin-derived components, each of which independently diffuses, and hence are distinct molecules. These formed associations in the operationally defined humic mixture. More-recent studies used liquid-state NMR on extracts in DMSO + 6% H_2SO_4 following exhaustive extractions in a base and base + urea. This 'humin' material was shown to contain lipoproteins, carbohydrates, peptidoglycans, aliphatic hydrocarbons, and some altered lignin residues. These components were held in intimate associations with the clay, and the oxidized lignin was considered to be trapped in the component mixture.

Emerging Concepts of Humic Structures

It is evident that the component molecules of HSs are still not unambiguously known. Considerable advances in identifying the organic structures have been made using NMR procedures, but so far the best identifications of the component molecules in the structures have been derived from identifying degradation products, and an awareness of the degradation mechanisms. We knew that 25~45% of the components of soil HAs are aromatic, and that value may be higher in some instances, e.g., for samples from the B_h horizons of some Podzols (Spodosols), and from soils containing significant amounts of char materials. Ten years ago, the consensus favored single-ring aromatic components with 3- to 5-ring substituents. That view has not changed, if the fused structures identified in pyrolysis are artifacts. However, fused ring structures are likely to be contained in HAs from soils where the burning of vegetation occurred, or where char materials were added to the soils. The hydrogen in 3 to 5 of the aromatic ring positions may be replaced by substituents consisting of hydroxyl, methoxyl, and aliphatic hydrocarbon structures. Some of these may be involved in linking aromatic structures. There is also evidence for aldehyde and ketones functional groups attached to some of the aromatic nuclei, for phenylpropane (3-carbon chains attached to the aromatic rings) units, and for hydroxyl and methoxyl substituents. The phenylpropane structures and hydroxy/methoxy substituents in the 3- and 4-, and in the 3-, 4-, and 5-ring positions would suggest origins in lignins, whereas the presence of these substituents in the 3- and 5-ring positions suggests origins in the skeletal structures or as metabolic products of microorganisms.

Fatty acids in degradation digests may be released from esters of phenols and other hydroxyl groups in the 'backbone' structures. Also these can arise from waxes and suberins and microbial products associated with HSs, and may contribute to the hydrophobic properties that have a role in the self-associations of HSs.

Titration data show that the acid groups in HSs provide a continuum of dissociable protons, and acid strengths ranging from strong to very weak. The strongest acids are carboxylic, and some are activated by neighboring groups. Phenolic hydroxyls also contribute to acidity, and these are most abundant in HA fractions. It would seem that their contribution to the total acidity is greatest in newly formed humic substances, especially in those with origins in the lignified components of plants. As oxidation takes place, the phenols are oxidized, and eventually carboxylic acids are formed. Some of the phenolic substituents can have enhanced acidity because of the influences of other substituents on the aromatic structures. Enols and other weakly dissociated groups also contribute to the charge characteristics under alkaline conditions.

The random coil model for solution conformations of HAs have served well for explaining interactions of chemicals with SOM components. However, evidence in recent times suggests that molecular associations give false high-MW properties to HSs, and that the major components have values of < 2000 Da. Numerous structures have been proposed for HSs during the course of the last few decades. In general, these were based on speculation, or on aspects of research that were in vogue at the time. Although it is not possible to give a definite structure for any of the major components of SOM, we are emerging from the world of "mystery and magic" where HSs are concerned, and there is real hope that soon there will be a realistic awareness of humic compositions and structures, and of the associations of molecules which give rise to the humic fractions. It is not necessary to know the exact structures in order to have a good understanding of the composition, or to be able to predict how the molecules will react with other species.

Soil pH

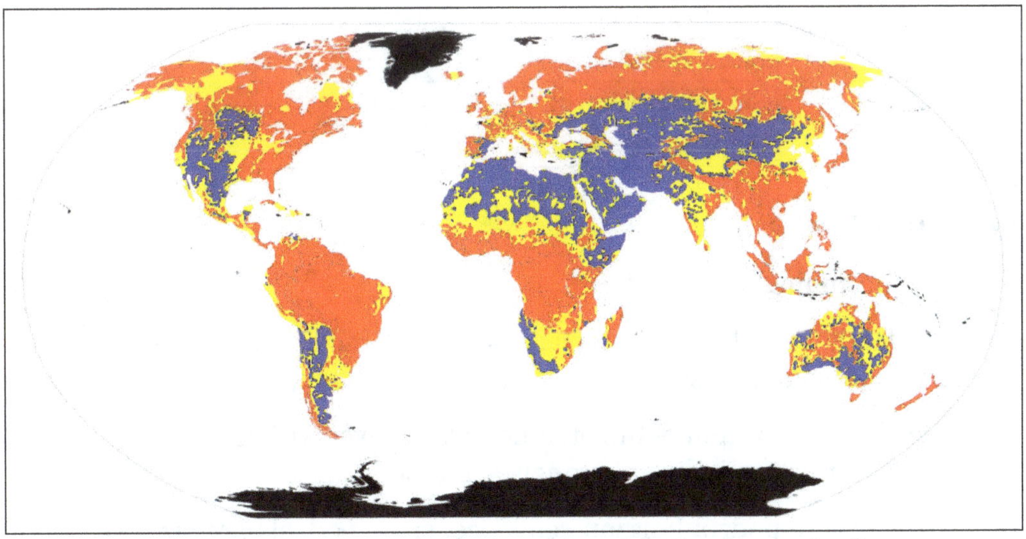

Global variation in soil pH. Red = acidic soil. Yellow = neutral soil.
Blue = alkaline soil. Black = no data.

Soil pH is a measure of the acidity or basicity (alkalinity) of a soil. pH is defined as the negative logarithm (base 10) of the activity of hydronium ions (H^+or, more precisely, $H_3O^+_{aq}$) in a solution. In soils, it is measured in a slurry of soil mixed with water (or a salt solution, such as 0.01 M $CaCl_2$), and normally falls between 3 and 10, with 7 being neutral. Acid soils have a pH below 7 and alkaline soils have a pH above 7. Ultra-acidic soils (pH < 3.5) and very strongly alkaline soils (pH > 9) are rare.

Soil pH is considered a master variable in soils as it affects many chemical processes. It specifically affects plant nutrient availability by controlling the chemical forms of the different nutrients and influencing the chemical reactions they undergo. The optimum pH range for most plants is between 5.5 and 7.5; however, many plants have adapted to thrive at pH values outside this range.

Classification of Soil pH Ranges

The United States Department of Agriculture Natural Resources Conservation Service classifies soil pH ranges as follows:

Denomination	pH range
Ultra acidic	< 3.5
Extremely acidic	3.5–4.4
Very strongly acidic	4.5–5.0
Strongly acidic	5.1–5.5
Moderately acidic	5.6–6.0
Slightly acidic	6.1–6.5
Neutral	6.6–7.3
Slightly alkaline	7.4–7.8
Moderately alkaline	7.9–8.4
Strongly alkaline	8.5–9.0
Very strongly alkaline	> 9.0

Determining pH

Methods of determining pH include:

- Observation of soil profile: Certain profile characteristics can be indicators of either acid, saline, or sodic conditions. Examples are:
    - Poor incorporation of the organic surface layer with the underlying mineral layer – this can indicate strongly acidic soils.
    - The classic podzol horizon sequence, since podzols are strongly acidic: In these soils, a pale eluvial (E) horizon lies under the organic surface layer and overlies a dark B horizon.
    - Presence of a caliche layer indicates the presence of calcium carbonates, which are present in alkaline conditions.
    - Columnar structure can be an indicator of sodic condition.

- Observation of predominant flora: Calcifuge plants (those that prefer an acidic soil) include Erica, Rhododendron and nearly all other Ericaceae species, many birch (Betula), foxglove (Digitalis), gorse (Ulex spp.), and Scots Pine (Pinus sylvestris). Calcicole (lime loving) plants include ash trees (Fraxinus spp.), honeysuckle (Lonicera), Buddleja, dogwoods (Cornus spp.), lilac (Syringa) and Clematis species.
- Use of an inexpensive pH testing kit, where in a small sample of soil is mixed with indicator solution which changes colour according to the acidity.
- Use of litmus paper: A small sample of soil is mixed with distilled water, into which a strip of litmus paper is inserted. If the soil is acidic the paper turns red, if basic, blue.
- Use of a commercially available electronic pH meter, in which a glass or solid-state electrode is inserted into moistened soil or a mixture (suspension) of soil and water; the pH is usually read on a digital display screen.
- Recently, spectrophotometric methods have been developed to measure soil pH involving addition of an indicator dye to the soil extract. These compared well to glass electrode measurements but offer substantial advantages such as lack of drift, liquid junction and suspension effects.

Precise, repeatable measures of soil pH are required for scientific research and monitoring. This generally entails laboratory analysis using a standard protocol; an example of such a protocol is that in the USDA Soil Survey Field and Laboratory Methods Manual. In this document the three-page protocol for soil pH measurement includes the following sections: Application; summary of method; interferences; safety; equipment; reagents; and procedure.

Factors Affecting Soil pH

The pH of a natural soil depends on the mineral composition of the parent material of the soil, and the weathering reactions undergone by that parent material. In warm, humid environments, soil acidification occurs over time as the products of weathering are leached by water moving laterally or downwards through the soil. In dry climates, however, soil weathering and leaching are less intense and soil pH is often neutral or alkaline.

Sources of Acidity

Many processes contribute to soil acidification. These include:

- Rainfall: Acid soils are most often found in areas of high rainfall. Rainwater has a slightly acidic pH (usually about 5.7) due to a reaction with CO_2 in the atmosphere that forms carbonic acid. When this water flows through soil it results in the leaching of basic cations from the soil as bicarbonates; this increases the percentage of Al^{3+} and H^+ relative to other cations.

- Root respiration and decomposition of organic matter by microorganisms releases CO_2 which increases the carbonic acid (H_2CO_3) concentration and subsequent leaching.
- Plant growth: Plants take up nutrients in the form of ions (e.g. NO^-_3, NH^+_4, Ca^{2+}, $H_2PO^-_4$), and they often take up more cations than anions. However plants must maintain a neutral charge in their roots. In order to compensate for the extra positive charge, they will release H^+ ions from the root. Some plants also exude organic acids into the soil to acidify the zone around their roots to help solubilize metal nutrients that are insoluble at neutral pH, such as iron (Fe).
- Fertilizer use: Ammonium (NH^+_4) fertilizers react in the soil by the process of nitrification to form nitrate (NO^-_3), and in the process release H^+ ions.
- Acid rain: The burning of fossil fuels releases oxides of sulfur and nitrogen into the atmosphere. These react with water in the atmosphere to form sulfuric and nitric acid in rain.
- Oxidative weathering: Oxidation of some primary minerals, especially sulfides and those containing Fe^{2+}, generate acidity. This process is often accelerated by human activity:
  - Mine spoil: Severely acidic conditions can form in soils near some mine spoils due to the oxidation of pyrite.
  - Acid sulfate soils formed naturally in waterlogged coastal and estuarine environments can become highly acidic when drained or excavated.

Total soil alkalinity increases with:

- Weathering of silicate, aluminosilicate and carbonate minerals containing Na^+, Ca^{2+}, Mg^{2+} and K^+.
- Addition of silicate, aluminosilicate and carbonate minerals to soils; this may happen by deposition of material eroded elsewhere by wind or water, or by mixing of the soil with less weathered material (such as the addition of limestone to acid soils).
- Addition of water containing dissolved bicarbonates (as occurs when irrigating with high-bicarbonate waters).

The accumulation of alkalinity in a soil (as carbonates and bicarbonates of Na, K, Ca and Mg) occurs when there is insufficient water flowing through the soils to leach soluble salts. This may be due to arid conditions, or poor internal soil drainage; in these situations most of the water that enters the soil is transpired (taken up by plants) or evaporates, rather than flowing through the soil.

The soil pH usually increases when the total alkalinity increases, but the balance of the added cations also has a marked effect on the soil pH. For example, increasing the amount of sodium in an alkaline soil tends to induce dissolution of calcium carbonate, which increases the pH. Calcareous soils may vary in pH from 7.0 to 9.5, depending on the degree to which Ca^{2+} or Na^{+} dominate the soluble cations.

Effect of Soil pH on Plant Growth

Acid Soils

Plants grown in acid soils can experience a variety of stresses including aluminium (Al), hydrogen (H), and/or manganese (Mn) toxicity, as well as nutrient deficiencies of calcium (Ca) and magnesium (Mg). Aluminium toxicity is the most widespread problem in acid soils. Aluminium is present in all soils, but dissolved Al^{3+} is toxic to plants; Al^{3+} is most soluble at low pH; above pH 5.0, there is little Al in soluble form in most soils. Aluminium is not a plant nutrient, and as such, is not actively taken up by the plants, but enters plant roots passively through osmosis. Aluminium inhibits root growth; lateral roots and root tips become thickened and roots lack fine branching; root tips may turn brown. In the root, the initial effect of Al^{3+} is the inhibition of the expansion of the cells of the rhizodermis, leading to their rupture; thereafter it is known to interfere with many physiological processes including the uptake and transport of calcium and other essential nutrients, cell division, cell wall formation, and enzyme activity.

Proton (H^{+} ion) stress can also limit plant growth. The proton pump, H^{+}-ATPase, of the plasmalemma of root cells works to maintain the near-neutral pH of their cytoplasm. A high proton activity (pH within the range 3.0–4.0 for most plant species) in the external growth medium overcomes the capacity of the cell to maintain the cytoplasmic pH and growth shuts down.

In soils with a high content of manganese-containing minerals, Mn toxicity can become a problem at pH 5.6 and lower. Manganese, like aluminium, becomes increasingly soluble as pH drops, and Mn toxicity symptoms can be seen at pH levels below 5.6. Manganese is an essential plant nutrient, so plants transport Mn into leaves. Classic symptoms of Mn toxicity are crinkling or cupping of leaves.

Nutrient Availability in Relation to Soil pH

Soil pH affects the availability of some plant nutrients:

- Aluminium toxicity has direct effects on plant growth; however, by limiting root growth, it also reduces the availability of plant nutrients. Because roots are damaged, nutrient uptake is reduced, and deficiencies of the macronutrients (nitrogen, phosphorus, potassium, calcium and magnesium) are frequently encountered in very strongly acidic to ultra-acidic soils ($pH<5.0$).
- Molybdenum availability is increased at higher pH; this is because the molybdate ion is more strongly sorbed by clay particles at lower pH.

- Zinc, iron, copper and manganese show decreased availability at higher pH (increased sorption at higher pH).
- The effect of pH on phosphorus availability varies considerably, depending on soil conditions and the crop in question. The prevailing view in the 1940s and 1950s was that P availability was maximized near neutrality (soil pH 6.5–7.5), and decreased at higher and lower pH. Interactions of phosphorus with pH in the moderately to slightly acidic range (pH 5.5–6.5) are, however, far more complex than is suggested by this view. Laboratory tests, glasshouse trials and field trials have indicated that increases in pH within this range may increase, decrease, or have no effect on P availability to plants.

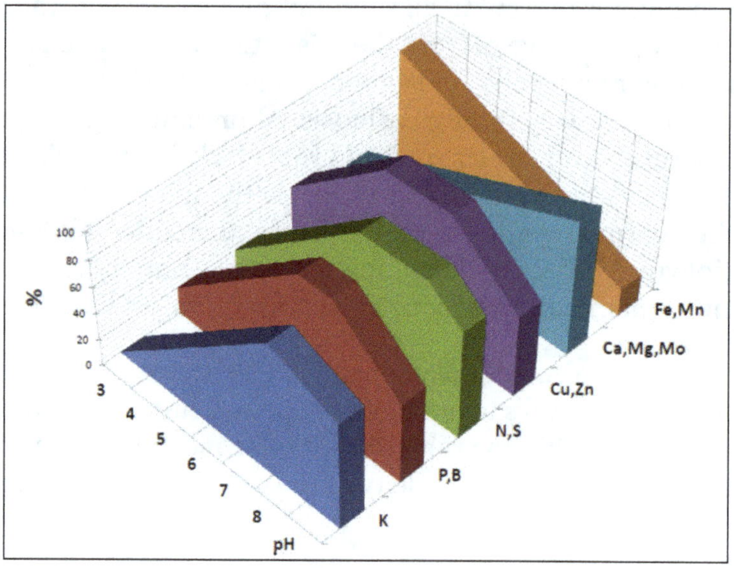

Nutrient availability in relation to soil pH.

Water Availability in Relation to Soil pH

Strongly alkaline soils are sodic and dispersive, with slow infiltration, low hydraulic conductivity and poor available water capacity. Plant growth is severely restricted because aeration is poor when the soil is wet; in dry conditions, plant-available water is rapidly depleted and the soils become hard and cloddy (high soil strength).

Many strongly acidic soils, on the other hand, have strong aggregation, good internal drainage, and good water-holding characteristics. However, for many plant species, aluminium toxicity severely limits root growth, and moisture stress can occur even when the soil is relatively moist.

Plant pH Preferences

In general terms, different plant species are adapted to soils of different pH ranges. For many species, the suitable soil pH range is fairly well known. Online databases of plant characteristics, such USDA PLANTS and Plants for a Future can be used to look up the suitable soil pH range of a wide range of plants. Documents like Ellenberg's indicator values for British plants can also be consulted.

However, a plant may be intolerant of a particular pH in some soils as a result of a particular mechanism, and that mechanism may not apply in other soils. For example, a soil low in molybdenum may not be suitable for soybean plants at pH 5.5, but soils with sufficient molybdenum allow optimal growth at that pH. Similarly, some calcifuges (plants intolerant of high-pH soils) can tolerate calcareous soils if sufficient phosphorus is supplied. Another confounding factor is that different varieties of the same species often have different suitable soil pH ranges. Plant breeders can use this to breed varieties that can tolerate conditions that are otherwise considered unsuitable for that species – examples are projects to breed aluminium-tolerant and manganese-tolerant varieties of cereal crops for food production in strongly acidic soils.

Changing Soil pH

Increasing pH of Acidic Soil

Finely ground agricultural lime is often applied to acid soils to increase soil pH (liming). The amount of limestone or chalk needed to change pH is determined by the mesh size of the lime (how finely it is ground) and the buffering capacity of the soil. A high mesh size (60 mesh = 0.25 mm; 100 mesh = 0.149 mm) indicates a finely ground lime that will react quickly with soil acidity. The buffering capacity of a soil depends on the clay content of the soil, the type of clay, and the amount of organic matter present, and may be related to the soil cation exchange capacity. Soils with high clay content will have a higher buffering capacity than soils with little clay, and soils with high organic matter will have a higher buffering capacity than those with low organic matter. Soils with higher buffering capacity require a greater amount of lime to achieve an equivalent change in pH.

Amendments other than agricultural lime that can be used to increase the pH of soil include wood ash, industrial calcium oxide (burnt lime), magnesium oxide, basic slag (calcium silicate), and oyster shells. These products increase the pH of soils through various acid-base reactions. Calcium silicate neutralizes active acidity in the soil by reacting with H^+ ions to form monosilicic acid (H_4SiO_4), a neutral solute.

Decreasing the pH of Alkaline Soil

The pH of an alkaline soil can be reduced by adding acidifying agents or acidic organic materials. Elemental sulfur (90–99% S) has been used at application rates of 300–500 kg/ha – it slowly oxidizes in soil to form sulfuric acid. Acidifying fertilizers, such as ammonium sulfate, ammonium nitrate and urea, can help to reduce the pH of a soil because ammonium oxidises to form nitric acid. Acidifying organic materials include peat or sphagnum peat moss.

However, in high-pH soils with a high calcium carbonate content (more than 2%), it can be very costly and/or ineffective to attempt to reduce the pH with acids. In such cases, it is often more efficient to add phosphorus, iron, manganese, copper and/or zinc instead, because deficiencies of these nutrients are the most common reasons for poor plant growth in calcareous soils.

Soil Carbon

Soil carbon refers to the solid terrestrial matter stored in global soils. This includes both soil organic matter and inorganic carbon as carbonate minerals. Soil carbon is a carbon sink in regard to the global carbon cycle, playing a role in biogeochemistry, climate change mitigation, and constructing global climate models. Soil carbon is present in two forms: inorganic and organic. Soil inorganic carbon consists of mineral forms of carbon, either from weathering of parent material, or from reaction of soil minerals with atmospheric CO_2. Carbonate minerals are the dominant form of soil carbon in desert climates. Soil organic carbon is present as soil organic matter. It includes relatively available carbon as fresh plant remains and relatively inert carbon in materials derived from plant remains: humus and charcoal.

Global Carbon Cycle

Although exact quantities are difficult to measure, human activities have caused substantial losses of soil organic carbon. Of the 2,700 Gt of carbon stored in soils worldwide, 1550 GtC is organic and 950 GtC is inorganic carbon, which is approximately three times greater than the current atmospheric carbon and 240 times higher compared with the current annual fossil fuel emission. The balance of soil carbon is held in peat and wetlands (150 GtC), and in plant litter at the soil surface (50 GtC). This compares to 780 GtC in the atmosphere, and 600 GtC in all living organisms. The oceanic pool accounts for 38,200 GtC. About 60 GtC/yr accumulates in the soil. This 60 GtC/yr is the balance of 120 GtC/yr contracted from the atmosphere by terrestrial plant photosynthesis reduced by 60 GtC/yr of plant respiration. An equivalent 60 GtC/yr is respired from soil, joining the 60G tC/yr plant respiration to return to the atmosphere.

Organic Carbon

Soil organic carbon is divided between living soil biota and dead biotic material derived from biomass. Together these comprise the soil food web, with the living component sustained by the biotic material component. Soil biota includes earthworms, nematodes, protozoa, fungi, bacteria and different arthropods.

Detritus resulting from plant senescence is the major source of soil organic carbon. Plant materials, with cell walls high in cellulose and lignin, are decomposed and the not-respired carbon is retained as humus. Cellulose and starches readily degrade, resulting in short residence times. More persistent forms of organic C include lignin, humus, organic matter encapsulated in soil aggregates, and charcoal. These resist alteration and have long residence times.

Soil organic carbon tends to be concentrated in the topsoil. Topsoil ranges from 0.5% to 3.0% organic carbon for most upland soils. Soils with less than 0.5% organic C are mostly limited to desert areas. Soils containing greater than 12 - 18% organic carbon are generally classified as organic soils. High levels of organic C develop in soils supporting wetland ecology, flood deposition, fire ecology, and human activity. Fire derived forms of carbon are present in most soils as unweathered charcoal and weathered black carbon. Soil organic carbon is typically 5 - 50% derived from char, with levels above 50% encountered in mollisol, chernozem, and terra preta soils.

Root exudates are another source of soil carbon. 5 - 20% of the total plant carbon fixed during photosynthesis is supplied as root exudates in support of rhizospheric mutualistic biota. Microbial populations are typically higher in the rhizosphere than in adjacent bulk soil.

Soil Health

A portable soil respiration system measuring soil CO_2 flux.

Organic carbon is vital to soil capacity to provide edaphic ecosystem services. The condition of this capacity is termed soil health, a term that communicates the value of understanding soil as a living system as opposed to an abiotic component. Specific carbon related benchmarks used to evaluate soil health include CO_2 release, humus levels, and microbial metabolic activity.

Losses

The exchange of carbon between soils and the atmosphere is a significant part of the world carbon cycle. Carbon, as it relates to the organic matter of soils, is a major component of soil and catchment health. Several factors affect the variation that exists in soil organic matter and soil carbon; the most significant has, in contemporary times, been the influence of humans and agricultural systems.

Although exact quantities are difficult to measure, human activities have caused massive losses of soil organic carbon. First was the use of fire, which removes soil cover and leads to immediate and continuing losses of soil organic carbon. Tillage and drainage both expose soil organic matter to oxygen and oxidation. In the Netherlands, East Anglia, Florida, and the California Delta, subsidence of peat lands from oxidation has been severe as a result of tillage and drainage. Grazing management that exposes soil (through either excessive or insufficient recovery periods) can also cause losses of soil organic.

Managing Soil Carbon

Natural variations in soil carbon occur as a result of climate, organisms, parent material, time, and relief. The greatest contemporary influence has been that of humans; for example, carbon in Australian agricultural soils may historically have been twice the present range that is typically from 1.6 to 4.6 per cent. It has long been encouraged that farmers adjust practices to maintain or increase the organic component in the soil. On one hand, practices that hasten oxidation of carbon (such as burning crop stubbles or over-cultivation) are discouraged; on the other hand, incorporation of organic material (such as in manuring) has been encouraged. Increasing soil carbon is not a straightforward matter; it is made complex by the relative activity of soil biota, which can consume and release carbon and are made more active by the addition of nitrogen fertilizers.

Data Available on Soil Organic Carbon

The most homogeneous and comprehensive data on the organic carbon/matter content of European soils remain those that can be extracted and/or derived from the European Soil Database in combination with associated databases on land cover, climate, and topography. The modelled data refer to carbon content (%) in the surface horizon of soils in Europe. In an inventory on available national datasets, seven member states of the European Union have available datasets on organic carbon. The LUCAS soil organic carbon data are measured surveyed points and the aggregated results at regional level show important findings. Finally, a new proposed model for estimation of soil organic carbon in agricultural soils has estimated current top SOC stock of 17.63 Gt in EU agricultural soils. This modelling framework has been updated by integrating the soil erosion component to estimate the lateral carbon fluxes.

Managing for Catchment Health

Much of the contemporary literature on soil carbon relates to its role, or potential, as an atmospheric carbon sink to offset climate change. Despite this emphasis, a much wider range of soil and catchment health aspects are improved as soil carbon is increased. These benefits are difficult to quantify, due to the complexity of natural resource systems and the interpretation of what constitutes soil health; nonetheless, several benefits are proposed in the following points:

- Reduced erosion, sedimentation: Increased soil aggregate stability means greater resistance to erosion; mass movement is less likely when soils are able to retain structural strength under greater moisture levels.
- Greater productivity: Healthier and more productive soils can contribute to positive socio-economic circumstances.
- Cleaner waterways, nutrients and turbidity: Nutrients and sediment tend to be retained by the soil rather than leach or wash off, and are so kept from waterways.
- Water balance: Greater soil water holding capacity reduces overland flow and recharge to groundwater; the water saved and held by the soil remains available for use by plants.
- Climate change: Soils have the ability to retain carbon that may otherwise exist as atmospheric CO_2 and contribute to global warming.
- Greater biodiversity: Soil organic matter contributes to the health of soil flora and, accordingly, the natural links with biodiversity in the greater biosphere.

Forest Soils

Forest soils constitute a large pool of carbon. Anthropogenic activities such as deforestation cause releases of carbon from this pool, which may significantly increase the concentration of greenhouse gas (GHG) in the atmosphere. Under the United Nations Framework Convention on Climate Change (UNFCCC), countries must estimate and report GHG emissions and removals, including changes in carbon stocks in all five pools (above- and below-ground biomass, dead wood, litter, and soil carbon) and associated emissions and removals from land use, land-use change and forestry activities, according to the Intergovernmental Panel on Climate Change's good practice guidance. Tropical deforestation represents nearly 25 percent of total anthropogenic GHG emissions worldwide. Deforestation, forest degradation, and changes in land management practices can cause releases of carbon from soil to the atmosphere. For these reasons, reliable estimates of soil organic carbon stock and stock changes are needed for Reducing emissions from deforestation and forest degradation and GHG reporting under the UNFCCC.

The government of Tanzania—together with the Food and Agriculture Organization of the United Nations and the financial support of the government of Finland—have implemented a forest soil carbon monitoring program to estimate soil carbon stock, using both survey and modelling-based methods.

Humus

In soil science, humus denominates the fraction of soil organic matter that is amorphous and without the "cellular cake structure characteristic of plants, micro-organisms or animals". Humus significantly affects the bulk density of soil and contributes to its retention of moisture and nutrients. In agriculture, "humus" sometimes also is used to describe mature or natural compost extracted from a woodland or other spontaneous source for use as a soil conditioner. It is also used to describe a topsoil horizon that contains organic matter (humus type, humus form, humus profile).

Humus is the dark organic matter that forms in soil when dead plant and animal matter decays. Humus has many nutrients that improve the health of soil, nitrogen being the most important. The ratio of carbon to nitrogen (C:N) of humus is 10:1.

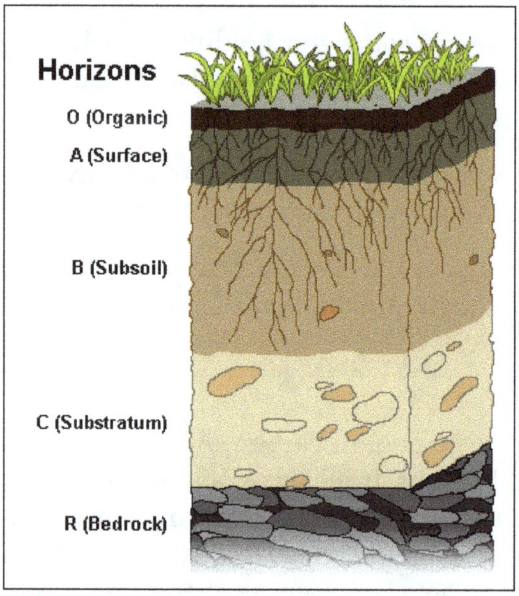

Humus has a characteristic black or dark brown color and is an accumulation of organic carbon. Besides the three major soil horizons of (A) surface/topsoil, (B) subsoil, and (C) substratum, some soils have an organic horizon (O) on the very surface. Hard bedrock (R) is not in a strict sense soil.

It is difficult to define humus precisely because it is a very complex substance which is not fully understood. Humus is different from decomposing soil organic matter. The latter looks rough and has visible remains of the original plant or animal matter. Fully humified humus, on the contrary, has a uniformly dark, spongy, and jelly-like appearance, and is amorphous; it may gradually decompose over several years or persist for millennia. It has no determinate shape, structure, or quality. However, when examined under a microscope, humus may reveal tiny plant, animal, or microbial remains that have been mechanically, but not chemically, degraded. This suggests an ambiguous boundary between humus and soil organic matter. While distinct, humus is an integral part of soil organic matter.

Humification

Microorganisms decompose a large portion of the soil organic matter into inorganic minerals that the roots of plants can absorb as nutrients. This process is termed "mineralization". In this process, nitrogen (nitrogen cycle) and the other nutrients (nutrient cycle) in the decomposed organic matter are recycled. Depending on the conditions in which the decomposition occurs, a fraction of the organic matter does not mineralize, and instead is transformed by a process called "humification" into concatenations of organic polymers. Because these organic polymers are resistant to the action of microorganisms, they are stable, and constitute humus. This stability implies that humus integrates into the permanent structure of the soil, thereby improving it.

Humification can occur naturally in soil or artificially in the production of compost. Organic matter is humified by a combination of saprotrophic fungi, bacteria, microbes and animals such as earthworms, nematodes, protozoa, and arthropods. Plant remains, including those that animals digested and excreted, contain organic compounds: sugars, starches, proteins, carbohydrates, lignins, waxes, resins, and organic acids. Decay in the soil begins with the decomposition of sugars and starches from carbohydrates, which decompose easily as detritivores initially invade the dead plant organs, while the remaining cellulose and lignin decompose more slowly. Simple proteins, organic acids, starches, and sugars decompose rapidly, while crude proteins, fats, waxes, and resins remain relatively unchanged for longer periods of time. Lignin, which is quickly transformed by white-rot fungi, is one of the primary precursors of humus, together with by-products of microbial and animal activity. The humus produced by humification is thus a mixture of compounds and complex biological chemicals of plant, animal, or microbial origin that has many functions and benefits in soil. Some judge earthworm humus (vermicompost) to be the optimal organic manure.

Stability

Much of the humus in most soils has persisted for more than 100 years, rather than having been decomposed into CO_2, and can be regarded as stable; this organic matter has been protected from decomposition by microbial or enzyme action because it is hidden (occluded) inside small aggregates of soil particles, or tightly sorbed or complexed to clays. Most humus that is not protected in this way is decomposed within 10 years and can be regarded as less stable or more labile. Stable humus contributes few plant-available nutrients in soil, but it helps maintain its physical structure. A very stable form of humus is formed from the slow oxidation of soil carbon after the incorporation of finely powdered charcoal into the topsoil. This process is speculated to have been important in the formation of the very fertile Amazonian terra preta do Indio.

Horizons

Humus has a characteristic black or dark brown color and is organic due to an accumulation of organic carbon. Soil scientists use the capital letters O, A, B, C, and E to identify the master horizons, and lowercase letters for distinctions of these horizons. Most soils have three major horizons: the surface horizon (A), the subsoil (B), and the substratum (C). Some soils have an organic horizon (O) on the surface, but this horizon can also be buried. The master horizon (E) is used for subsurface horizons that have significantly lost minerals (eluviation). Bedrock, which is not soil, uses the letter R.

Benefits of Soil Organic Matter and Humus

The importance of chemically stable humus is thought by some to be the fertility it provides to soils in both a physical and chemical sense, though some agricultural experts put a greater focus on other features of it, such as its ability to suppress disease. It helps the soil retain moisture by increasing microporosity, and encourages the formation of good soil structure. The incorporation of oxygen into large organic molecular assemblages generates many active, negatively charged sites that bind to positively charged ions (cations) of plant nutrients, making them more available to the plant by way of ion exchange. Humus allows soil organisms to feed and reproduce, and is often described as the "life-force" of the soil.

- The process that converts soil organic matter into humus feeds the population of microorganisms and other creatures in the soil, and thus maintains high and healthy levels of soil life.
- The rate at which soil organic matter is converted into humus promotes (when fast) or limits (when slow) the coexistence of plants, animals, and microorganisms in the soil.
- Effective humus and stable humus are additional sources of nutrients for microbes: the former provides a readily available supply and the latter acts as a longeval storage reservoir.
- Decomposition of dead plant material causes complex organic compounds to be slowly oxidized (lignin-like humus) or to decompose into simpler forms (sugars and amino sugars, and aliphatic and phenolic organic acids), which are further transformed into microbial biomass (microbial humus) or reorganized, and further oxidized, into humic assemblages (fulvic acids and humic acids), which bind to clay minerals and metal hydroxides. The ability of plants to absorb humic substances with their roots and metabolize them has been long debated. There is now a consensus that humus functions hormonally rather than simply nutritionally in plant physiology.
- Humus is a colloidal substance and increases the cation exchange capacity of soil, hence its ability to store nutrients by chelation. While these nutrient

cations are available to plants, they are held in the soil and prevented from being leached by rain or irrigation.

- Humus can hold the equivalent of 80–90% of its weight in moisture, and therefore increases the soil's capacity to withstand drought.
- The biochemical structure of humus enables it to moderate, i.e. buffer, excessive acidic or alkaline soil conditions.
- During humification, microbes secrete sticky, gum-like mucilages; these contribute to the crumby structure (tilth) of the soil by adhering particles together and allowing greater aeration of the soil. Toxic substances such as heavy metals and excess nutrients can be chelated, i. e., bound to the organic molecules of humus, and so prevented from leaching away.
- The dark, usually brown or black, color of humus helps to warm cold soils in Spring.

Soil Contamination

Excavation showing soil contamination at a disused gasworks.

Soil contamination or soil pollution as part of land degradation is caused by the presence of xenobiotics (human-made) chemicals or other alteration in the natural soil environment. It is typically caused by industrial activity, agricultural chemicals or improper disposal of waste. The most common chemicals involved are petroleum hydrocarbons, polynuclear aromatic hydrocarbons (such as naphthalene and benzo(a)pyrene), solvents, pesticides, lead, and other heavy metals. Contamination is correlated with the degree of industrialization and intensity of chemical substance. The concern over soil contamination stems primarily from health risks, from direct contact with the contaminated soil, vapours from the contaminants, or from secondary contamination of water supplies within and underlying the soil. Mapping of contaminated soil sites and the resulting cleanups are time consuming and expensive tasks, requiring extensive amounts of geology, hydrology, chemistry, computer modeling skills, and GIS in Environmental Contamination, as well as an appreciation of the history of industrial chemistry.

In North America and Western Europe the extent of contaminated land is best known, with many of countries in these areas having a legal framework to identify and deal with this environmental problem. Developing countries tend to be less tightly regulated despite some of them having undergone significant industrialization.

Causes of Soil Pollution

Soil pollution can be caused by the following (non-exhaustive list):

- Microplastics.
- Oil spills.
- Mining and activities by other heavy industries.
- Accidental spills may happen during activities, etc.
- Corrosion of underground storage tanks (including piping used to transmit the contents).
- Acid rain.
- Intensive farming.
- Agrochemicals, such as pesticides, herbicides and fertilizers.
- Petrochemicals.
- Industrial accidents.
- Road debris.
- Drainage of contaminated surface water into the soil.
- Ammunitions, chemical agents, and other agents of war.
- Waste disposal:
    - Oil and fuel dumping.
    - Nuclear wastes.
    - Direct discharge of industrial wastes to the soil.
    - Discharge of sewage.
    - Landfill and illegal dumping.
    - Coal ash.
    - Electronic waste.

The most common chemicals involved are petroleum hydrocarbons, solvents, pesticides, lead, and other heavy metals.

Any activity that leads to other forms of soil degradation (erosion, compaction, etc.) may indirectly worsen the contamination effects in that soil remediation becomes more tedious. Historical deposition of coal ash used for residential, commercial, and industrial heating, as well as for industrial processes such as ore smelting, were a common source of contamination in areas that were industrialized before about 1960. Coal naturally concentrates lead and zinc during its formation, as well as other heavy metals to a lesser degree. When the coal is burned, most of these metals become concentrated in the ash (the principal exception being mercury). Coal ash and slag may contain sufficient lead to qualify as a "characteristic hazardous waste", defined in the US as containing more than 5 mg/l of extractable lead using the TCLP procedure. In addition to lead, coal ash typically contains variable but significant concentrations of polynuclear aromatic hydrocarbons (PAHs; e.g., benzo(a)anthracene, benzo(b)fluoranthene, benzo(k)fluoranthene, benzo(a)pyrene, indeno(cd)pyrene, phenanthrene, anthracene, and others). These PAHs are known human carcinogens and the acceptable concentrations of them in soil are typically around 1 mg/kg. Coal ash and slag can be recognised by the presence of off-white grains in soil, gray heterogeneous soil, or (coal slag) bubbly, vesicular pebble-sized grains.

Treated sewage sludge, known in the industry as biosolids, has become controversial as a "fertilizer". As it is the byproduct of sewage treatment, it generally contains more contaminants such as organisms, pesticides, and heavy metals than other soil. In the European Union, the Urban Waste Water Treatment Directive allows sewage sludge to be sprayed onto land. The volume is expected to double to 185,000 tons of dry solids in 2005. This has good agricultural properties due to the high nitrogen and phosphate content. In 1990/1991, 13% wet weight was sprayed onto 0.13% of the land; however, this is expected to rise 15 fold by 2005. Advocates say there is a need to control this so that pathogenic microorganisms do not get into water courses and to ensure that there is no accumulation of heavy metals in the top soil.

Pesticides and Herbicides

A pesticide is a substance used to kill a pest. A pesticide may be a chemical substance, biological agent (such as a virus or bacteria), antimicrobial, disinfectant or device used against any pest. Pests include insects, plant pathogens, weeds, mollusks, birds, mammals, fish, nematodes (roundworms) and microbes that compete with humans for food, destroy property, spread or are a vector for disease or cause a nuisance. Although there are benefits to the use of pesticides, there are also drawbacks, such as potential toxicity to humans and other organisms.

Herbicides are used to kill weeds, especially on pavements and railways. They are similar to auxins and most are biodegradable by soil bacteria. However, one group derived from trinitrotoluene (2:4 D and 2:4:5 T) have the impurity dioxin, which is very toxic and causes fatality even in low concentrations. Another herbicide is Paraquat. It is highly toxic but it rapidly degrades in soil due to the action of bacteria and does not kill soil fauna.

Insecticides are used to rid farms of pests which damage crops. The insects damage not only standing crops but also stored ones and in the tropics it is reckoned that one third of the total production is lost during food storage. As with fungicides, the first insecticides used in the nineteenth century were inorganic e.g. Paris Green and other compounds of arsenic. Nicotine has also been used since the late eighteenth century.

There are now two main groups of synthetic insecticides:

- Organochlorines include DDT, Aldrin, Dieldrin and BHC. They are cheap to produce, potent and persistent. DDT was used on a massive scale from the 1930s, with a peak of 72,000 tonnes used 1970. Then usage fell as the harmful environmental effects were realized. It was found worldwide in fish and birds and was even discovered in the snow in the Antarctic. It is only slightly soluble in water but is very soluble in the bloodstream. It affects the nervous and endocrine systems and causes the eggshells of birds to lack calcium causing them to be easily breakable. It is thought to be responsible for the decline of the numbers of birds of prey like ospreys and peregrine falcons in the 1950s – they are now recovering. As well as increased concentration via the food chain, it is known to enter via permeable membranes, so fish get it through their gills. As it has low water solubility, it tends to stay at the water surface, so organisms that live there are most affected. DDT found in fish that formed part of the human food chain caused concern, but the levels found in the liver, kidney and brain tissues was less than 1 ppm and in fat was 10 ppm, which was below the level likely to cause harm. However, DDT was banned in the UK and the United States to stop the further buildup of it in the food chain. U.S. manufacturers continued to sell DDT to developing countries, who could not afford the expensive replacement chemicals and who did not have such stringent regulations governing the use of pesticides.
- Organophosphates, e.g. parathion, methyl parathion and about 40 other insecticides are available nationally. Parathion is highly toxic, methyl-parathion is less so and Malathion is generally considered safe as it has low toxicity and is rapidly broken down in the mammalian liver. This group works by preventing normal nerve transmission as cholinesterase is prevented from breaking down the transmitter substance acetylcholine, resulting in uncontrolled muscle movements.

Agents of War

The disposal of munitions, and a lack of care in manufacture of munitions caused by the urgency of production, can contaminate soil for extended periods. There is little published evidence on this type of contamination largely because of restrictions placed by governments of many countries on the publication of material related to war effort.

However, mustard gas stored during World War II has contaminated some sites for up to 50 years and the testing of Anthrax as a potential biological weapon contaminated the whole island of Gruinard.

Health Effects

Contaminated or polluted soil directly affects human health through direct contact with soil or via inhalation of soil contaminants which have vaporized; potentially greater threats are posed by the infiltration of soil contamination into groundwater aquifers used for human consumption, sometimes in areas apparently far removed from any apparent source of above ground contamination. This tends to result in the development of pollution-related diseases.

Health consequences from exposure to soil contamination vary greatly depending on pollutant type, pathway of attack and vulnerability of the exposed population. Chronic exposure to chromium, lead and other metals, petroleum, solvents, and many pesticide and herbicide formulations can be carcinogenic, can cause congenital disorders, or can cause other chronic health conditions. Industrial or man-made concentrations of naturally occurring substances, such as nitrate and ammonia associated with livestock manure from agricultural operations, have also been identified as health hazards in soil and groundwater.

Chronic exposure to benzene at sufficient concentrations is known to be associated with higher incidence of leukemia. Mercury and cyclodienes are known to induce higher incidences of kidney damage and some irreversible diseases. PCBs and cyclodienes are linked to liver toxicity. Organophosphates and carbonates can induce a chain of responses leading to neuromuscular blockage. Many chlorinated solvents induce liver changes, kidney changes and depression of the central nervous system. There is an entire spectrum of further health effects such as headache, nausea, fatigue, eye irritation and skin rash for the above cited and other chemicals. At sufficient dosages a large number of soil contaminants can cause death by exposure via direct contact, inhalation or ingestion of contaminants in groundwater contaminated through soil.

The Scottish Government has commissioned the Institute of Occupational Medicine to undertake a review of methods to assess risk to human health from contaminated land. The overall aim of the project is to work up guidance that should be useful to Scottish Local Authorities in assessing whether sites represent a significant possibility of significant harm (SPOSH) to human health. It is envisaged that the output of the project will be a short document providing high level guidance on health risk assessment with reference to existing published guidance and methodologies that have been identified as being particularly relevant and helpful. The project will examine how policy guidelines have been developed for determining the acceptability of risks to human health and propose an approach for assessing what constitutes unacceptable risk in line with the criteria for SPOSH as defined in the legislation and the Scottish Statutory Guidance.

Ecosystem Effects

Not unexpectedly, soil contaminants can have significant deleterious consequences for ecosystems. There are radical soil chemistry changes which can arise from the presence of many hazardous chemicals even at low concentration of the contaminant species. These changes can manifest in the alteration of metabolism of endemic microorganisms and arthropods resident in a given soil environment. The result can be virtual eradication of some of the primary food chain, which in turn could have major consequences for predator or consumer species. Even if the chemical effect on lower life forms is small, the lower pyramid levels of the food chain may ingest alien chemicals, which normally become more concentrated for each consuming rung of the food chain. Many of these effects are now well known, such as the concentration of persistent DDT materials for avian consumers, leading to weakening of egg shells, increased chick mortality and potential extinction of species.

Effects occur to agricultural lands which have certain types of soil contamination. Contaminants typically alter plant metabolism, often causing a reduction in crop yields. This has a secondary effect upon soil conservation, since the languishing crops cannot shield the Earth's soil from erosion. Some of these chemical contaminants have long half-lives and in other cases derivative chemicals are formed from decay of primary soil contaminants.

Cleanup Options

Cleanup or environmental remediation is analyzed by environmental scientists who utilize field measurement of soil chemicals and also apply computer models (GIS in Environmental Contamination) for analyzing transport and fate of soil chemicals. Various technologies have been developed for remediation of oil-contaminated soil and sediments. There are several principal strategies for remediation:

- Excavate soil and take it to a disposal site away from ready pathways for human or sensitive ecosystem contact. This technique also applies to dredging of bay muds containing toxins.
- Aeration of soils at the contaminated site (with attendant risk of creating air pollution).
- Thermal remediation by introduction of heat to raise subsurface temperatures sufficiently high to volatize chemical contaminants out of the soil for vapor extraction. Technologies include ISTD, electrical resistance heating (ERH), and ET-DSP.
- Bioremediation, involving microbial digestion of certain organic chemicals. Techniques used in bioremediation include landfarming, biostimulation and bioaugmentating soil biota with commercially available microflora.

- Extraction of groundwater or soil vapor with an active electromechanical system, with subsequent stripping of the contaminants from the extract.
- Containment of the soil contaminants (such as by capping or paving over in place).
- Phytoremediation, or using plants (such as willow) to extract heavy metals.
- Mycoremediation, or using fungus to metabolize contaminants and accumulate heavy metals.
- Remediation of oil contaminated sediments with self-collapsing air microbubbles.
- Surfactant leaching.

Soil Acidification

Soil acidification is the buildup of hydrogen cations, which reduces the soil pH. Chemically, this happens when a proton donor gets added to the soil. The donor can be an acid, such as nitric acid, sulfuric acid, or carbonic acid. It can also be a compound such as aluminium sulfate, which reacts in the soil to release protons. Acidification also occurs when base cations such as calcium, magnesium, potassium and sodium are leached from the soil.

Soil acidification naturally occurs as lichens and algae begin to break down rock surfaces. Acids continue with this dissolution as soil develops. With time and weathering, soils become more acidic in natural ecosystems. Soil acidification rates can vary, and increase with certain factors such as acid rain, agriculture, and pollution.

Causes of Soil Acidification

- Acid Rain: Rainfall is naturally acidic due to carbonic acid forming from carbon dioxide in the atmosphere. This compound causes rainfall pH to be around 5.0-5.5. When rainfall has a lower pH than natural levels, it can cause rapid acidification of soil. Sulfur dioxide and nitrogen oxides are precursors of stronger acids that can lead to acid rain production when they react with water in the atmosphere. These gases may be present in the atmosphere due to natural sources such as lightning and volcanic eruptions, or from anthropogenic emissions. Basic cations like calcium are leached from the soil as acidic rainfall flows, which allows aluminum and proton levels to increase.
- Biological weathering: Plant roots acidify soil by releasing protons and organic acids so as to chemically weather soil minerals. Decaying remains of dead plants on soil may also form organic acids which contribute to soil acidification.

Acidification from leaf litter on soil is more pronounced under coniferous trees such as pine, spruce and fir, which return fewer base cations to the soil,rather than under deciduous trees.

- Parent Materials: Certain parent materials also contribute to soil acidification. Granites and their allied igneous rocks are called "acidic" because they have a lot of free quartz, which produces silicic acid on weathering. Also, they have relatively low amounts of calcium and magnesium. Some sedimentary rocks such as shale and coal are rich in sulfides, which, when hydrated and oxidized, produce sulfuric acid which is much stronger than silicic acid. Many coal soils are too acidic to support vigorous plant growth, and coal gives off strong precursors to acid rain when it is burned. Marine clays are also sulfide-rich in many cases, and such clays become very acidic if they are drained to an oxidizing state.
- Soil Amendments: Soil amendments such as fertilizers and manures can cause soil acidification. Sulfur based fertilizers can be highly acidifying, examples include elemental sulfur and iron sulfate while others like potassium sulfate have no significant effect on soil pH. While most nitrogen fertilizers have an acidifying effect, ammonium-based nitrogen fertilizers are more acidifying than other nitrogen sources. Ammonia-based nitrogen fertilizers include ammonium sulfate, diammonium phosphate, monoammonium phosphate, and ammonium nitrate. Organic nitrogen sources, such as urea and compost, are less acidifying. Nitrate sources which have little or no ammonium, such as calcium nitrate, magnesium nitrate, potassium nitrate, and sodium nitrate, are not acidifying.
- Pollution: Acidification may also occur from nitrogen emissions into the air, as the nitrogen may end up deposited into the soil. Animal livestock is responsible for nearly 65 percent of man-made ammonia emissions.

  Anthropogenic sources of sulfur dioxides and nitrogen oxides play a major role in increase of acid rain production. The use of fossil fuels and motor exhaust are the largest anthropogenic contributors to sulfuric gases and nitrogen oxides, respectively.

Effects

Soil acidification can cause damage to plants and organisms in the soil. In plants, soil acidification results in smaller, less durable roots. Acidic soils sometimes damage the root tips reducing further growth. Plant height is impaired and seed germination also decreases. Soil acidification impacts plant health, resulting in reduced cover and lower plant density. Overall,stunted growth is seen in plants. Soil acidification is directly linked to a decline in endangered species of plants.

In the soil, acidification reduces microbial and macrofaunal diversity. This can reduce soil structure decline which makes it more sensitive to erosion. There are less nutrients available in the soil, larger impact of toxic elements to plants, and consequences to soil biological functions (such as nitrogen fixation). At a larger scale, soil acidification is linked to losses in agricultural productivity due to these effects.

Prevention and Management

Soil acidification is a common issue in long-term crop production which can be reduced by lime application. In soybean and corn crops grown in acidic soils, lime application resulted in nutrient restoration, increase in soil pH, increase in root biomass, and better plant health.

Different management strategies may also be applied to prevent further acidification: using less acidifying fertilizers, considering fertilizer amount and application timing to reduce nitrate-nitrogen leaching, good irrigation management with acid-neutralizing water, and considering the ratio of basic nutrients to nitrogen in harvested crops. Sulfur fertilizers should only be used in responsive crops, with a high rate of crop recovery.

Through reduction of anthropogenic sources of sulfur dioxides, nitrogen oxides, and with air-pollution control measures, let us try to reduce acid rain and soil acidification all over the world.

Microorganisms in Soil

Soil microorganisms can be grouped into bacteria, actinomycetes, fungi, algae, protozoa, and nematodes. Apart from the dead plant or animal residues in soils, SOM is composed of a significant content of living microorganisms and their dead fractions. The humus fraction is resistant to microbial decomposition and persists for thousands of years contributing to the long-lived carbon pool in soils. Soil microorganisms are involved in the decomposition of soil organic matter, and the rate of decomposition depends both on the nature of microorganisms in soil and the nature of organic matter sources. Enhancing the activities of soil fungi has been recognized as one of the potential options for reducing SOC turnover, thereby increasing carbon sequestration. Melanin, chitin, and glomalin are examples of fungal-derived recalcitrant residues that tend to exist for a long time in soils. Apart from the humification process, soil microorganisms are involved in mineralization of SOM, thereby resulting in the loss of carbon from soils. The priming effect (PE) is referred to as the short-term change in the turnover of native SOM. It often occurs after inputs such as organic amendments, mineral fertilizers, and root exudates. The PE can be either positive, whereby the addition of carbon increases the decomposition of native soil carbon, or negative whereby the decomposition of native soil carbon is reduced after the input of organic amendments. Soil microorganisms contribute to the PE, thereby impacting soil carbon stabilization. Carbon use efficiency (CUE) is used to understand soil microbial metabolism. Organisms with a lower CUE respire a higher proportion of metabolized carbon as CO_2. CUE determination in soils is controversial, and some researchers have reported bacterial CUE to be lower than fungal CUE while others have found the opposite. Soil houses billions of both bacteria and fungi, and, therefore, it is difficult to differentiate CUE from each group; however, it has been reported that a greater proportion of carbon is stored in fungal-dominated soil systems than bacterial dominated soil systems. Soil microbes indirectly influence the physically protected SOM by improving soil aggregation, thereby enhancing carbon stabilization in soils.

Microorganisms are very small forms of life that can sometimes live as single cells, although many also form colonies of cells. A microscope is usually needed to see individual cells of these organisms. Many more microorganisms exist in topsoil, where food sources are plentiful, than in subsoil. They are especially abundant in the area immediately next to plant roots (called the rhizosphere), where sloughed-off cells and chemicals released by roots provide ready food sources. These organisms are primary decomposers of organic matter, but they do other things, such as provide nitrogen through fixation to help growing plants, detoxify harmful chemicals (toxins), suppress disease organisms, and produce products that might stimulate plant growth. Soil microorganisms have had another direct importance for humans—they are the source of most of the antibiotic medicines we use to fight diseases.

Bacteria

Bacteria live in almost any habitat. They are found inside the digestive system of animals, in the ocean and fresh water, in compost piles (even at temperatures over 130 °F), and in soils. Although some kinds of bacteria live in flooded soils without oxygen, most require well-aerated soils. In general, bacteria tend to do better in neutral pH soils than in acid soils.

In addition to being among the first organisms to begin decomposing residues in the soil, bacteria benefit plants by increasing nutrient availability. For example, many bacteria dissolve phosphorus, making it more available for plants to use. Bacteria are also very helpful in providing nitrogen to plants, which they need in large amounts but is often deficient in agricultural soils. You may wonder how soils can be deficient in nitrogen when we are surrounded by it—78% of the air we breathe is composed of nitrogen gas. Yet plants as well as animals face a dilemma similar to that of the Ancient Mariner, who was adrift at sea without fresh water: "Water, water, everywhere nor any drop to drink." Unfortunately, neither animals nor plants can use nitrogen gas (N_2) for their nutrition. However, some types of bacteria are able to take nitrogen gas from the atmosphere and convert it into a form that plants can use to make amino acids and proteins. This conversion process is known as nitrogen fixation.

Some nitrogen-fixing bacteria form mutually beneficial associations with plants. One such symbiotic relationship that is very important to agriculture involves the nitrogen-fixing rhizobia group of bacteria that live inside nodules formed on the roots of legumes. These bacteria provide nitrogen in a form that leguminous plants can use, while the legume provides the bacteria with sugars for energy.

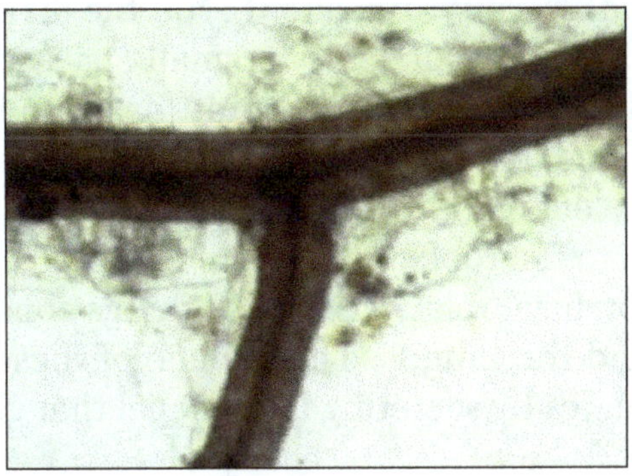

Root heavily infected with mycorrhizal fungi.

People eat some legumes or their products, such as peas, dry beans, and tofu made from soybeans. Soybeans, alfalfa, and clover are used for animal feed. Clovers and hairy vetch are grown as cover crops to enrich the soil with organic matter, as well as nitrogen, for the following crop. In an alfalfa field, the bacteria may fix hundreds of pounds of nitrogen per acre each year. With peas, the amount of nitrogen fixed is much lower, around 30 to 50 pounds per acre.

The actinomycetes, another group of bacteria, break large lignin molecules into smaller sizes. Lignin is a large and complex molecule found in plant tissue, especially stems that is difficult for most organisms to break down. Lignin also frequently protects other molecules like cellulose from decomposition. Actinomycetes have some characteristics similar to those of fungi, but they are sometimes grouped by themselves and given equal billing with bacteria and fungi.

Fungi

Fungi are another type of soil microorganism. Yeast is a fungus used in baking and in the production of alcohol. Other fungi produce a number of antibiotics. We have all probably let a loaf of bread sit around too long only to find fungus growing on it. We have seen or eaten mushrooms, the fruiting structures of some fungi. Farmers know that fungi cause many plant diseases, such as downy mildew, damping-off, various types of root rot, and apple scab. Fungi also initiate the decomposition of fresh organic residues. They help get things going by softening organic debris and making it easier for other organisms to join in the decomposition process. Fungi are also the main decomposers of lignin and are less sensitive to acid soil conditions than bacteria. None are able to function without oxygen. Low soil disturbance resulting from reduced tillage systems tends to promote organic residue accumulation at and near the surface. This tends to promote fungal growth, as happens in many natural undisturbed ecosystems.

Many plants develop a beneficial relationship with fungi that increases the contact of roots with the soil. Fungi infect the roots and send out root-like structures called hyphae. The hyphae of these mycorrhizal fungi take up water and nutrients that can then feed the plant. The hyphae are very thin, about 1/60 the diameter of a plant root, and are able to exploit the water and nutrients in small spaces in the soil that might be inaccessible to roots. This is especially important for phosphorus nutrition of plants in low-phosphorus soils. The hyphae help the plant absorb water and nutrients, and in return the fungi receive energy in the form of sugars, which the plant produces in its leaves and sends down to the roots. This symbiotic interdependency between fungi and roots is called a mycorrhizal relationship. All things considered, it's a pretty good deal for both the plant and the fungus. The hyphae of these fungi help develop and stabilize larger soil aggregates by secreting a sticky gel that glues mineral and organic particles together.

Algae

Algae, like crop plants, convert sunlight into complex molecules like sugars, which they can use for energy and to help build other molecules they need. Algae are found in abundance in the flooded soils of swamps and rice paddies, and they can be found on the surface of poorly drained soils and in wet depressions. Algae may also occur in relatively dry soils, and they form mutually beneficial relationships with other organisms. Lichens found on rocks are an association between a fungus and an alga.

Protozoa

Protozoa are single-celled animals that use a variety of means to move about in the soil. Like bacteria and many fungi, they can be seen only with the help of a microscope. They are mainly secondary consumers of organic materials, feeding on bacteria, fungi, other protozoa, and organic molecules dissolved in the soil water. Protozoa—through their grazing on nitrogen-rich organisms and excreting wastes—are believed to be responsible for mineralizing (releasing nutrients from organic molecules) much of the nitrogen in agricultural soils.

Biogeochemical Processes at Soil-root Interface

Rhizosphere is the narrow zone of soil influenced by the roots and exudates. The extent of the rhizosphere, which is usually considered to extend from the root surface out into the soil for a few millimeters, may vary with plant species, age of the plant, soil properties, the level of physical, chemical, and biological stress and so on. The biological and chemical processes which occur at soil-root interface are dramatically different from those occurring in the bulk soil due to the interaction among plant roots, microorganisms and soil components. More intense microbial activity and larger microbial population occur in this zone than in bulk soil, due to the release by roots of large amounts of organic compounds. Microbial population in the rhizosphere can be 10-100 times larger than the population in bulk soil due to the much higher amount of organic substances present at soil-plant interface. A multitude of compounds are released into the rhizosphere by plants and microorganisms. Rhizosphere C flow has been estimated to account for a major fraction, up to 30-40% of plant primary production. Root exudates comprise both high- and low-molecular-weight substances released by the roots. The most important high-molecular-weight compounds are mucilage, polysaccharides, and ectoenzymes, whereas the main constituents of the low-molecular-weight root exudates are carbohydrates, organic acids, aminoacids, peptides and, phenolics. The relative abundance of organic acids is aliphatic acids > aromatic acids > amino acids. The hydroxyl-, di-, and tricarboxylic acids (e.g., oxalic, malic, succinic, citric, tartaric, and malonic acids) are the most abundant aliphatic acids present in the rhizosphere. Diverse bacteria and fungi produce large amounts of siderophores and some plants release phytosiderophores. These compounds, which are diamino-carboxylic acids, may be excreted into the rhizosphere and might have an important role in the mobilization and uptake of Fe. In addition to having a high affinity for Fe(III), they can also bind other metals (Cd, Zn, Cu). Root products that are not C compounds are relatively few (protons, inorganic ions, water, electrons).

Biomolecules produced by plants and microorganisms as well soil organic substances are involved in many processes at soil-root interface such as: a) Weathering of clay minerals and release of Al and Fe (as well as Ca, Zn, Cu, Mn) b) formation of nanoprecipitates and organo-mineral complexes, c) sorption/desorption of cations and anions on/from soil colloids, d) bioavailability of nutrients and pollutants, and e) complexation and detoxification of toxic elements.

Interactions between Biomolecules and Clay Minerals

Formation and Nature of Nanoprecipitates at Soil-root Interface

Rhizosphere soil exhibits differences in physico-chemical properties and mineralogy compared with bulk soil. It is a major active microsite where interactions among roots, symbiotic microorganisms, bacteria, soil constituents (minerals, organic substances), and soil solution take place and are involved in solubilization and insolubilization processes and the formation of secondary minerals that affect the structure, texture, and reactivity of soil. As summarized by Séguin *et al.* "the rhizosphere is considered to be a microenvironment that is more acidic than the bulk soil, which contain a greater amount and diversity of organic matter, mainly through rhizodeposition. The soil under root influence generally has a lower pH, although the effect on pH depends on the nutritional status of the plant. A plant absorbs both anions and cations, but rarely in an equivalent amount. In order to maintain its electroneutrality and that of the surrounding soil solution, the plant must either exsude OH^- or H^+ to equilibrate the anion-cation balance. It can then be expected that mineral weathering will be increased in the rhizosphere".

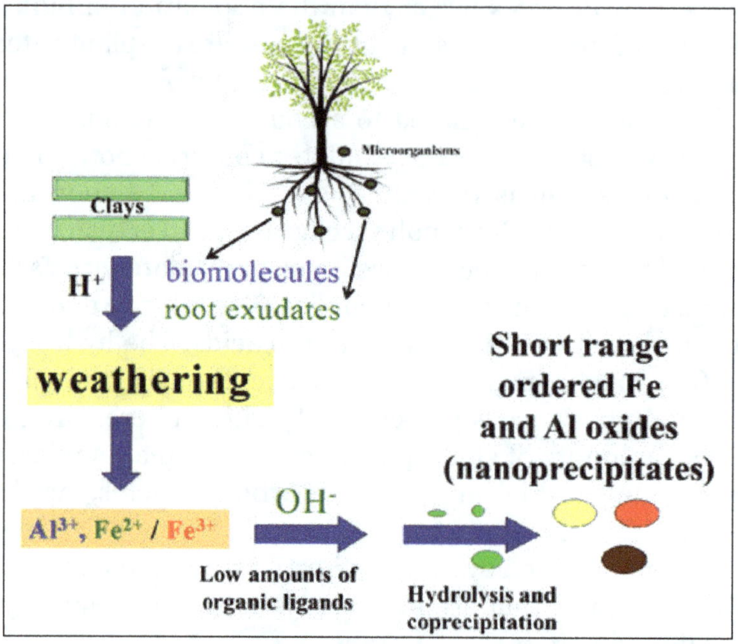

Interaction between root exudates and biomolecules released by plants and microorganisms. Chelating organic ligands promote weathering of clay minerals and the release of Al and Fe. Organic ligands coprecipitate with Al and Fe, promoting the formation of short-range ordered Al- and Fe-oxides with particles extremely small in size (< 100 nm; nanoprecipitates).

Chelating root exudates and organic compounds released by microorganisms as well as organic acids are involved in the weathering of minerals and in the subsequent release and transformations of Fe and Al at root-soil interface. Their solubilisation capabilities depend on their chemical composition, nature of clay minerals and the mineral element considered. In particular, the ability of (phyto)siderophores to form stable complexes with Fe (complex stability constant range from 10^{23} to 10^{52}), Al and other cations suggests that these chelating ligands may influence metal mobility in the rhizosphere by affecting mineral weathering. Such processes of acid dissolution and complexation of minerals due to metabolites are considered as indirect weathering processes. Some bacteria, including anaerobic species and fungi can attack mineral surfaces and thereby mobilize constituents of the mineral. One of the modes of attack of minerals includes indirect attack with a metabolically produced ligand that forms a highly soluble product with a mineral component.

Aluminum and Fe are released to soil solution through chemical and biochemical processes, undergoes hydrolysis and may give rise to precipitation products. Clay minerals, pH, organic ligands are the most important factors which influence the release and phase transformation of Al and Fe at soil-plant interface and the mineralogy, order, particle size, specific surface area and reactivity of the metal precipitation products towards nutrients and pollutants. Organic compounds which form stable complexes with Fe and Al ions, compete with OH^- ions and strongly perturb the hydrolytic reactions, retarding or inhibiting the crystallization of Fe- and Al-(oxy)hydroxides through occupation of Fe-and Al-(oxy)hydroxides coordination sites. As a consequence, the formation of noncrystalline or short-range ordered Al and Fe precipitates (e.g. ferrihydrite) is promoted at soil-root interface. This behavior is similar to that ascertained in the weathering interface between lichens and rocks.

The increased weathering of minerals in the rhizosphere induces changes in the abundance and the forms of metals at soil-root interface, as compared to the bulk soil. Greater contents of acid ammonium oxalate (AAO) extractable Al and Fe have been found in the rhizosphere than in the bulk soil. AAO solutions are used to extract the inorganic noncrystalline Al and Fe oxides together with organically complexed metals most of which occur as weathering products in soils. The greater abundance of noncrystalline oxides is then the result of an increase in weathering rates.

Many laboratory studies have been carried out on the influence of strongly chelating organic ligands on the nature, size of the particles and crystallinity of Al and Fe precipitates. These researches have demonstrated that organic chelating anions (e.g. citric, oxalic, malic, tartaric, acid) have a tremendous influence on Al and Fe nucleation reactions, their transformations and surface properties of the final precipitation products. These ligands cause structural distortion of these precipitates and, thus, increase their specific surface and reactive sites. It has been demonstrated that in the absence of ligands well crystallized Fe or Al oxides (gibbsite, goethite, hematite, lepidocrocite) forms in few days or weeks at $pH > 4.0$, whereas at certain concentrations of strongly chelating ligands (L/Al or Fe molar ratio ranging from 0.01 to 0.1) short range ordered precipitates (ferrihydrites, noncrystalline Fe and Al oxides, poorly crystalline boehmite [AlOOH]) form and remain unchanged for years. X-ray diffraction analysis and electron microscope studies showed that some precipitates were noncrystalline even after a long aging period (more than 10 years at 20-50 °C) and usually appeared to be agglomerates of spherical particles, whose size was < 50 nm. Other studies demonstrated that the nature and concentration of organic ligands also affected the morphology of short-range ordered precipitates. Poorly crystalline boehmite (short-range ordered Al-oxyhydroxides, AlOOH) formed under certain ionic conditions and showed fibrous morphology, but there were differences in the degree of fiber development depending on the ligand/Al molar ratio, pH, time of aging and temperature. Many researchers have demonstrated that poorly crystalline and noncrystalline Al- and Fe-oxides are very important in stabilizing organic substances.

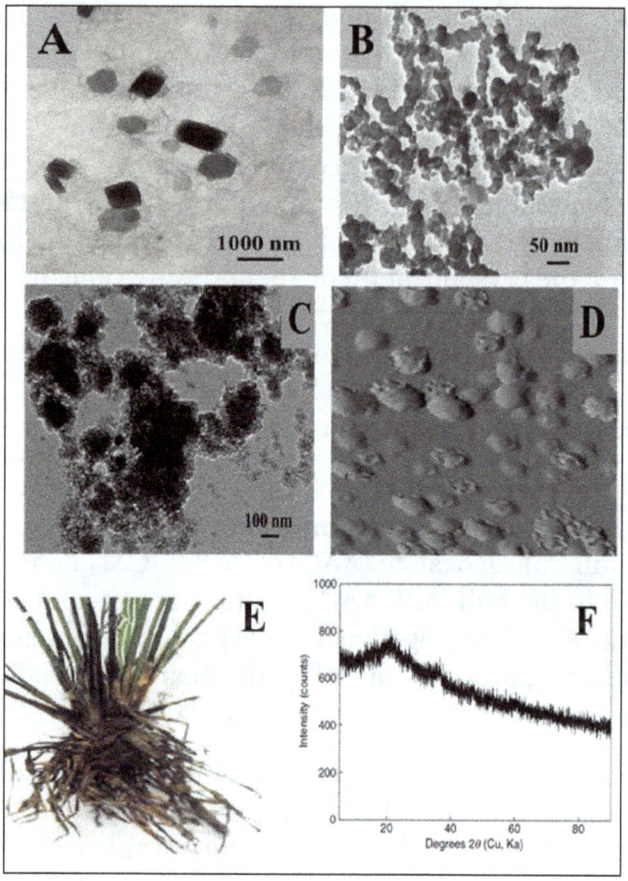

Transmission electron micrographs of (A) gibbsite formed at pH 6 in the absence of organic ligands, (B) noncrystalline Al oxides formed at pH 5.0 at tannate/Al molar ratio of 0.1 aged 9 years at 50 °C, (C) ferrihydrite formed at pH 7.0 and citrate/Fe molar ratio of 0.1 aged 1 year, (D) an atomic force micrograph of the same sample (C); (E) iron plaques on plant roots, (F) 2-line ferrihydrite in root plaques of rice seedlings detected by XRD.

Biomolecules interact with Fe released from minerals promoting the formation of poorly crystalline ferrihydrite. The degree of ordering of ferrihydrite is variable and a range of X-ray diffraction patterns may be obtained. The two extremes of ordering are referred as 2-line (0.252 and 0.148 nm) and 6-line ferrihydrite (additional lines at 0.225, 0.197, 0.172, and 0.151 nm), because the XRD patterns range from two to six reflections as structural order increases. It is well known that Fe precipitates, commonly called iron plaques, form on the roots of many wetland plants. Numerous studies suggest that the predominant minerals present in Fe root plaques are Fe(III)-oxyhydroxides, including ferrihydrite (mainly), goethite and lepidocrocite, but also Mn oxides may form. Iron plaques often contain high concentrations of metals. A possible role of iron plaque is to facilitate the uptake and translocation of nutrients and microelements (e.g., Cd), or may act as a barrier reducing the uptake of potentially phytotoxic metals and metalloids (As) by plants. Xu and Yu found that the iron plaque coatings the rice seedling roots appeared to be to XRD 2-line ferrihydrite, but noncrystalline Fe precipitates were probably present being the peaks barely visible.

Interaction between Abiotic and Biotic Soil Components

Root exudates [low molecular mass organic ligands (LMMOLs) as well as biopolymers (e.g. proteins (enzymes), nucleotides, polysaccharides, lipids] interact with clay minerals, OH-Al and OH-Fe species and/or microorganisms forming organo-mineral complexes. Microorganisms-mineral complexes, which form by electrostatic or hydrophobic interaction between microorganisms and minerals are other common complexes at soil-plant interface. Microorganisms participate in the formation of organo-mineral complexes in which these complexes provide substrates and protection. They contribute to the formation of organo-mineral complexes excreting polysaccharides and proteins.

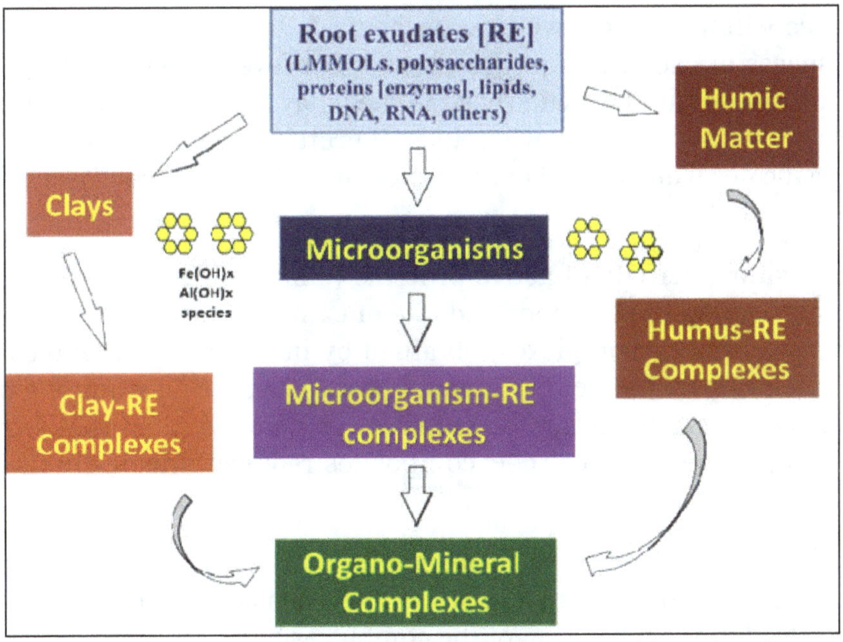

Interactions of root exudates with soil components (clay minerals, Fe(Al)(OH)x species, humic matter, microorganisms) forming organo-mineral complexes of different complexity.

Because of the complexity of natural organo-mineral complexes, many studies have been carried out on the nature, reactivity and activity (if active biopolymers [e.g. enzymes] were present) of synthetic complexes obtained by the interaction among two or more soil components (abiotic and biotic). Short-range order minerals, OH-Al and OH-Fe species are particularly important in stabilizing organic substances in soil environments, because usually organic molecules (LMMOLs, (phyto)siderophores as well as biopolymers) are strongly fixed on these sorbents forming inner-sphere complexes. Organic matter associated with (oxalate-extractable) Fe and Al is potentially stable for thousands of years.

Organic compounds are sorbed on the surfaces of phyllosilicates or even intercalated into the interlayers of expandable clays (montmorillonite, vermiculite) only at low pH (say pH < 5) where they do not present a negative charge. However, polyvalent cations but mainly Al-OH and Fe-OH species, facilitate the sorption of organic substances onto phyllosilicates. It has been demonstrated that monomers, polymers and solid species of Al and Fe can react with organic compounds forming complexes which may present a positive or neutral charge and, then, may be easily sorbed on the external negative surfaces of phyllosilicates and may be intercalated into the interlayers of montmorillonite even in neutral or alkaline environments. Many studies have also been carried out on the interaction between biopolymers mainly proteins [enzymes] with phyllosilicates and/or humic acids (or other organic substances). Proteins may be easily adsorbed and intercalated in the interlayers of montmorillonite, but the amounts of proteic molecules sorbed and intercalated are affected by pH, size and isoelectric point of proteins. Much evidence seems to demonstrate that large biopolymers or organomineral complexes with a net positive or even neutral charge may be intercalated into the lamellar spaces of expansible phyllosilicates. However, when biopolymers or organomineral complexes are particularly large they may be sorbed only on the external surfaces of clay minerals. Only small amounts of proteins may be adsorbed on the surfaces of phyllosilicates when the pH of the system is higher than the isoelectric point of the proteic molecules.

Sorption and residual activity of active proteins (enzymes) is of paramount importance for understanding the behavior and fate of extracellular proteinic molecules in the rhizosphere. Synthetic complexes, obtained by immobilizing enzymes onto clay minerals, humic acids or humic-clay composites, were used to study the properties of soil-bound enzymes. Numerous researches have demonstrated that enzymes immobilized on soil components or their composites reduced drastically their activity (up to 80%).

The nature, physico-chemical properties and reactivity of organomineral complexes are affected not only by the nature and amount of each soil component, both abiotic and biotic, involved in the formation of the complexes but also by the sequence of addition of the components involved in their formation. Studies have been carried out on the influence of the sequence of addition of montmorillonite, hydroxyl-Al species and humic acids, tannic acid or proteins. The different sequence addition of OH-Al(Fe) species, montmorillonite and large organic ligands promote the formation of complexes with different d-spacing and surface properties, which are attributed to differences in the nature of OH-Al-organomineral complexes and different interlayering of OH-Al species and biomolecules. Some researchers studied the influence of the sequence addition of montmorillonite, hydroxy-Al species and humic or tannic acids on the nature of organo-clay complexes formed at pH 4.5-7.0. Negligible amounts of humic or tannic acids were held on the clay surfaces in the absence of Al, whereas in the presence of Al, hydroxyl-Al-humate or tannate complexes were easily sorbed on clay surfaces.

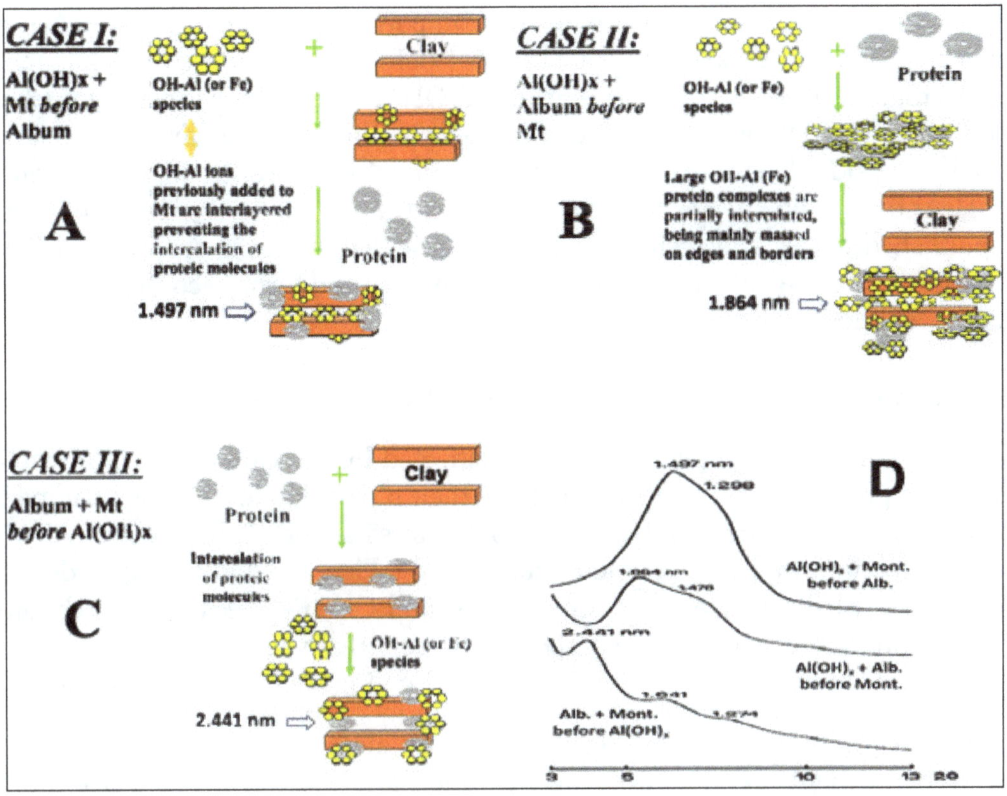

Schematic representation of the sorption of albumin onto montmorillonite (Mt) in the Al(OH)x-Mt complexes, obtained by differently mixing Al(OH)x species, albumin and montmorillonite in order to have three diverse systems: Al(OH)x + Mt before protein (Case I, A), Al(OH)x + protein before Mt (Case II, B), and protein + Mt before Al(OH) x (Case III, C). (D) X-ray diffraction patterns of the complexes described above (A-C).

Their distribution on the external surface and in the interlayer space of montmorillonite was a consequence of how the components reacted with each other. The final complexes also showed distinct differences in cation exchange capacity (CEC), carbon content, extractable Al, titratable acidity, and mode of aggregation after drying. Later, De Cristofaro and Violante studied the effect of the sequence of addition of OH-Al species, protein molecules (albumin) and montmorillonite as well as pH on the nature and surface properties of the final protein-OH-Al-montmorillonite complexes. They demonstrated that OH-Al species initially added to the clay were previously interlayered and consequently prevented the intercalation of the proteinic molecules, which were sorbed on the external surfaces (Case I: Al(OH)x + Mt before Alb system). In contrast, when OH-Al species and proteins were added to the clay as a mixture, large OH-Al-protein polymers formed, which were only partially intercalated into the interlamellar spaces of the clay (Case II: Al(OH)x + Alb before Mt system). Finally, when albumin was added to montmorillonite before OH-Al species (Case III: Alb + Mt before Al(OH)x system) the proteinic molecules were able to penetrate easily into the interlayers of montmorillonite. X-ray diffraction patterns showed that intercalation of albumin occurred more easily in the Alb + Mt before Al(OH)x system (d-spacing of 2.441 nm) than A(OH)xl + Mt before Alb and Al(OH)x + Alb before Mt systems (d-spacings of 1.864 and 1.497 nm, respectively) in the order listed. When an enzyme (urease) was used the residual activity of complexes obtained by differently mixing the enzyme with OH-Al or OH-Fe species and clay minerals was substantially different (Violante, unpublished data). The residual activity of selected enzymes onto phyllosilicates, Al[Fe]-oxides, tannate or humic acids and their binary and ternary complexes has been extensively studied.

Sorption and Desorption Processes at Soil-plant Interface

Mobility of elements at soil-plant interface is affected by sorption-desorption reactions, which are the predominant factors that control their bioavailability. The soil components responsible for the sorption of nutrients and pollutants include phyllosilicates, soil organic matter, carbonates, variable charge minerals (crystalline and short-range ordered Fe-, Al-, Mn-oxides, phyllosilicates coated by OH–Al and OH–Fe species), microorganisms and organomineral complexes. Soil components differ greatly in their sorption capacities, their cation and anion exchange capacities, and the binding energies of their sorption sites.

Two surface complexes exist and are described by the configuration geometry of the sorbate at the sorbent surface. These include inner- and outer-sphere surface complexes. Outer-sphere complexes involve electrostatic coulombic interactions, and are thus weak compared with inner-sphere and are reversible. Inner- and outer-sphere complexations can occur simultaneously. Specific sorption is characterized for higher selectivity and stronger binding (weaker reversibility), while nonspecific sorption is known for lower selectivity and greater reversibility. The strength of the sorption of cations and anions is an important factor governing their mobility in soils.

Sorption of elements onto soil components is influenced by many factors as pH, nature of the sorbents, redox reactions, presence of foreign ions. Briefly, cations show typical ion exchange behavior on clay minerals. Surface bonding is prevalently electrostatic, but polyvalent cations (e.g. Zn, Cu, Cd, Co, Ca) may be strongly adsorbed on the edge sites of phyllosilicates due to the presence of –SiOH or –AlOH groups capable of chemisorbing these ions. Organic matter and variable charge minerals (metal oxides, allophanes, imogolite) are much more effective scavenger of polyvalent cations, because complexation processes are the dominant binding mechanisms. Polyvalent cations are usually sorbed on these sorbents forming inner-sphere complexes.

Evidence on the sorption of metal cations on microorganisms has been reported. Free-living bacteria and their extracellular macromolecular products (e.g. fibrils) can accumulate metal ions and may have mineral coatings with bound metals on their surfaces. The cell surfaces of all bacteria are largely negatively charged, containing different types of negatively charged functional groups, such as carboxyl, hydroxyl, and phosphoryl that can sorb metal cations. Biosorption comprises a variety of processes, including ion exchange, chelation, sorption, and diffusion through cell walls and membranes all of which are dependent on the species used, the biomass origin, and solution chemistry.

Anions sorb primarily to variable charge minerals (metal oxides and noncrystalline aluminum silicates, allophanes), carbonate, and at the edges of phyllosilicates. Usually, they are not sorbed on soil organic matter, but certain elements (e.g. borate, arsenate, arsenite) are found to be bound to humic substances.

Sorption of anions onto variable charge minerals and soils varies with pH. With increasing pH values, within a certain range, sorption decreases (due to a decrease in the positive charge of minerals) or else increases to a maximum close to the pKa for anions of monoprotic conjugate acids and then decreases. Ligands that are specifically sorbed, forming *inner-sphere* complexes, replace $-OH^-$ or $-OH_2$ groups from the surfaces of variable charge minerals. Ligands, which form *inner-sphere* complexes, such as phosphate, arsenate, arsenite (on Fe-oxides), molybdate, selenite, and polydentate chelating low-molecular mass organic ligands (LMMOLs, e.g. oxalate, citrate, tartrate, malate) or soil organic substances, may form different surface complexes on inorganic soil components: Monodentate, bidentate-binuclear and bidentate-mononuclear complexes in different proportions depending on the pH and surface coverage.

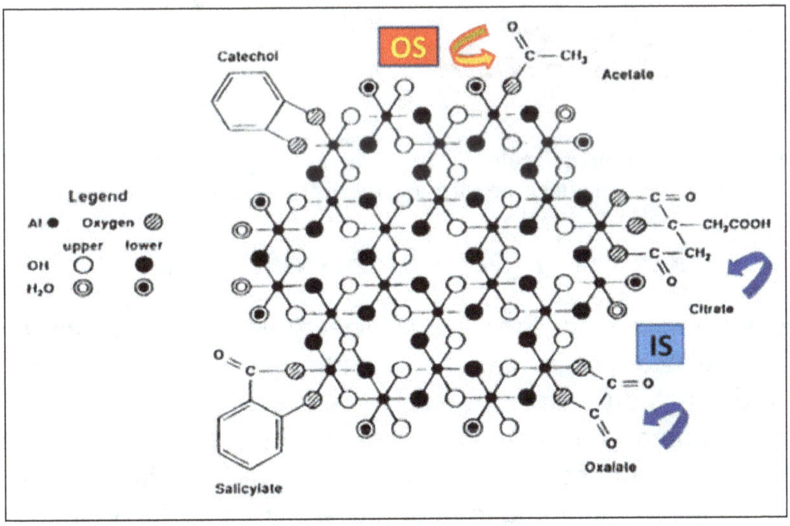

Examples of interactions between selected root exudates and an Al-oxide, forming monodentate, bidentate or tridentate complexes (inner-sphere [IS] or outer-sphere [OS]).

Sulfate and selenate seem to be able to form only outer-sphere complexes. However, there is some spectroscopic evidence that sulfate can also be sorbed as an inner-sphere complex. Recent studies have demonstrated that sulfate may be sorbed onto the surfaces of variable charge minerals, forming both inner-sphere and outer-sphere complexes, with the former occurring mainly at pH << 6.0 and increasing sulfate concentrations. The halide anions, with the exception of fluoride, and organic monocarboxylic (acetate) anions bond by outer-sphere electrostatic attraction and are sorbed only if variable charge minerals surfaces are positively charged (pH < PZC of the sorbents).

Role of Biomolecules on the Mobility of Elements at Soil-root Interface

Root exudates and organic substances released by microorganisms (mainly LMMOLs) play a vital role on the sorption/desorption of elements on/from soil components and organo-mineral complexes and then on their availability for plants and soil organisms. In figure below a model of exudates flux in the rhizosphere is depicted and its influence on the mobility of nutrients and pollutants. Sorption/desorption processes of cations are different from those of anions.

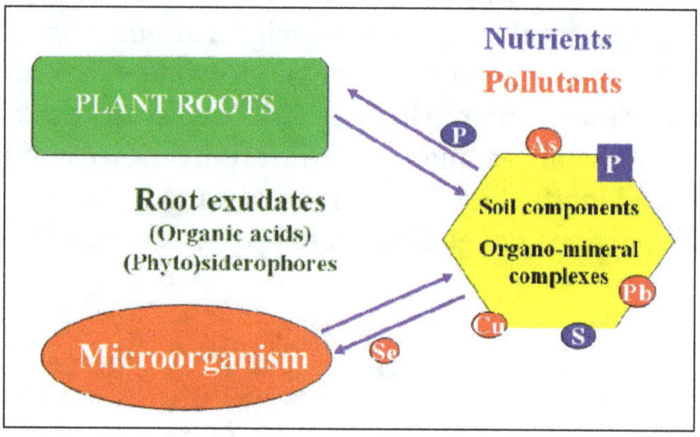

A possible model of biomolecules (root exudates and [phyto]siderophores) flux in the rhizosphere,which affect the sorption/desorption of nutrients and pollutants.

Effects of Biomolecules on the Mobility of Cations

Sorption of elements in cationic form onto soil components may be either enhanced or inhibited by the presence of organic acids (as oxalic, citric, malic, tartaric, malonic acid) and aminoacids through different processes, including alteration of surface charge, site competition, formation of different complexes in solution, ternary complex formed on the surfaces of soil components. Ternary surfaces complexes may be classified as type A, in which the metal ion is bonded to both the sorbent and the organic ligand, or type B, in which the ligand is bonded to the sorbent between the surface and the metal ion. The structure of sorbed species determines their stability and the extent of metal ion partitioning from solution to solid phase.

Organic ligands usually inhibit the sorption of cations on negatively charged sorbents, as phyllosilicates, organic matter and microorganisms, by forming stable soluble negatively charged complexes with polyvalent cations. Some researchers demonstrated that various LMMOLs inhibited at different degrees the sorption of cations onto phyllosilicates. The influence of chelating organic ligands on the sorption of polyvalent cations onto soil organic matter and organo-mineral complexes has received attention. Usually, they compete with organic substances and reduce cation sorption. Wu *et al.* found that the amount of Pb sorbed onto soil organic matter decreased with increasing concentration of EDTA, clearly because EDTA very strongly complexes with Pb ions and prevents their fixation on soil organic matter.

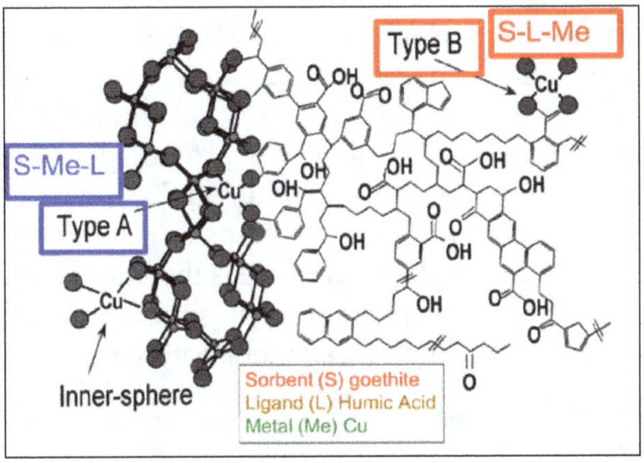

Demonstration of some possible bonding configurations of Cu (Me) on goethite (S)–organic matter (L) complexes: Inner-sphere complexation of Cu at goethite sites, and type-A- (S-Me-L) and B-(S-L-Me) ternary complexes.

The sorption of polyvalent cations in the presence of biomolecules onto the surfaces of variable charge minerals (iron and aluminum oxides, allophanes, imogolite) may be affected by several processes, which may occur simultaneously: (i) Increased retention of metal cations for the decrease of the positive charge of the mineral when a ligand is sorbed, (ii) increased retention due to the formation of negatively charged complexes between a metal and a ligand, which have a higher affinity for the sorbent, (iii) decreased retention due to the block of sorption sites by foreign ligands on the minerals, (iv) decreased retention due to the strong complexation of metal in solution in nonsorbing form, with lower free metal activity leading to desorption of surface-bound metal, (v) dissolution of the mineral in the presence of relatively high concentration of chelating organic ligands, leading to lesser metal retention. In each of the cases described before, the effect of a ligand on the sorption of metal cations depends not only on the strengths of the interactions between metal ion, ligand, and sorbent but also on their concentrations. Certainly, the initial ligand/metal molar ratio, the nature of the ligand and the metal, the surface properties of the sorbent seem to be critical in determining whether the sorption of a cation is enhanced or inhibited.

Further, because anionic ligands are usually sorbed more strongly at low pH, whereas metal cations are fixed more easily at high pH, the same ligand may enhance the sorption of a given metal under some conditions but may suppress it under others. The formation of sorbent–metal–ligand (type A) or sorbent–ligand–metal (type B) ternary complexes has been suggested as the main mechanisms also for the promotive effects of organic ligands on the sorption of metal cation onto variable charge minerals.

Effect of (Phyto)siderophores on the Sorption of Cations

(Phyto)siderophores are low molecular weight organic ligands secreted by aerobic microorganisms and some plants, to acquire Fe. They can also form stable complexes with other metals, such as Zn, Cu, Cd, Pb, Al, Ga, Pd. These biomolecules have an important impact on cations mobility in the rhizosphere. They have a behavior different from that of strongly chelating organic acids. Neubauer *et al.* studied the influence of desferrioxamine B(DFO-B) on the sorption of Cu, Zn and Cd onto montmorillonite and kaolinite. In acidic systems and in the absence of DFO-B only small amounts of these elements were sorbed on phyllosilicates. The presence of DFO-B promoted the sorption of Cu, Zn and Cd on montmorillonite, palygorskite and sepiolite. Studies on the sorption of rare earth elements and actinides on Pseudomonas fluorescens cells also demonstrated that in the presence of DFO-B the sorption of these elements substantially increases. Electrostatic interactions between the negative surfaces of phyllosilicates or bacterial cells and DFO-B-Me complexes are the reasons for enhanced metal cation sorption onto these sorbents.

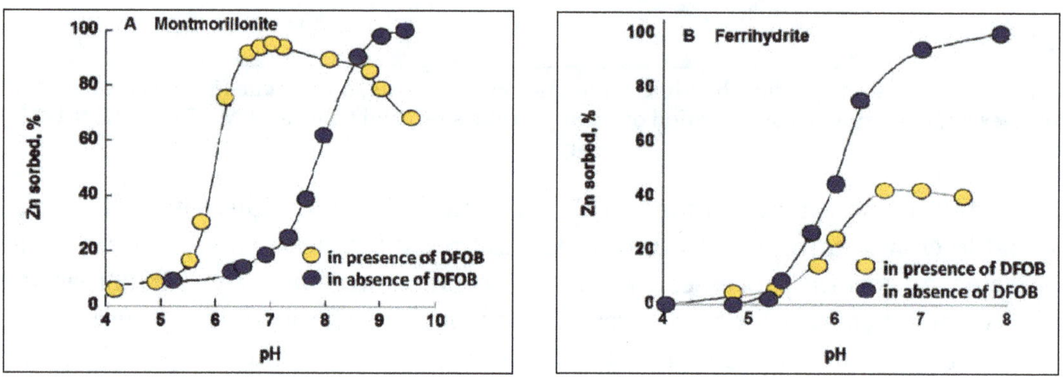

Sorption of Zn as a function of pH in the absence and presence of the hydroxamatesiderophore DFOB on (A) montmorillonite (Zn concentration 87 μmol/L) and (B) ferrihydrite (Zn concentration 10 μmol/L).

The sorption of Cu, Zn, Pb, Eu(III) on variable charge minerals (ferrihydrite, boehmite, goethite) in the presence of DFO-Bwas different from that onto montmorillonite or bacterial cells, which present negative sorption sites. Their sorption on these sorbents was strongly prevented, because positively charged DFO-B-Me complexes are not attracted by the positively charged surfaces of ferrihydrite, goethite and boehmite.

The effect of four different trihydroxamate siderophores (including DFO-B) on Pb and Cd sorption to kaolinite at pH 4.5 to 9.0 was also studied by Hepinstall et al. These researchers demonstrated that at pH > 6.0 all of the studied ligands decreased Pb sorption (from 5 to 75%), the greater effects were at higher pH. The same siderophores decreased Cd sorption onto kaolinite at pH > 8.0. This behavior must be ascribed to the fact that kaolinite presents much more variable sites than montmorillonite. Anyway, these researchers also observed a slight increase in sorption of Pb from pH 5.0-6.0 in the presence of DFO-B onto kaolinite probably due to sorption of the doubly positive charged $H_3Pb(DFO\text{-}B)^{2+}$ on the permanent negative sites of this mineral. Hepinstall *et al.* concluded that siderophores have the ability to either enhance or inhibit metal mobility at soil-root interface and that further researches are necessary in order to understand the factors (molecular structure and concentration of different siderophores, nature of the sorbents, pH, and nature of heavy metals) which affect the mobility of polyvalent cations.

Several plant species are able to release organic acids from their roots in response to toxic elements (e.g. monomers of Al) present in the rooting medium while others release these compounds when growing under a deficient supply of iron or phosphate. Some of them are involved in energy production as intermediates in the tricarboxylic acid cycle (e.g. citrate, malate, oxalate, tartrate), while others are directly or indirectly involved in many other metabolic processes. Because these organic acids are strongly chelating agents, they can form strong complexes with Al and heavy metals thereby reducing their activity in solution and toxicity to roots.

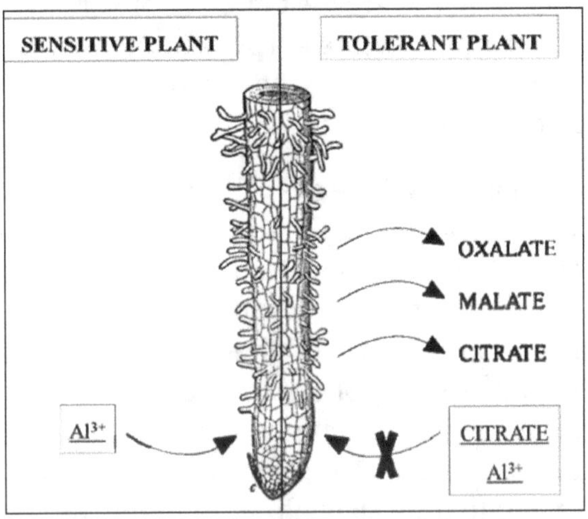

Tolerant plants release large amounts of chelating organic ligands (oxalate, malate, citrate, tartrate) which complex Al species which become less toxic and do not enter into the plant roots.

Effect of Root Exudates on the Sorption of Anions

The effect of root exudates on the sorption of nutrients and toxic elements in anionic form (phosphate, sulfate, selenite/selenate, arsenate/arsenite, molibdate) has received great attention because the supply of these anions to plants should be strongly influenced by the nature and concentration of root exudates. The competition between anions (e.g. LMMOLs and phosphate or sulfate) is not only competition for the occupancy of sites available on the sorbents but it also involves a change in the electric potential of the surfaces due of the sorbents.

Among the first studies on the influence of LMMOLs on phosphate sorption are those of Deb and Datta and Nagarajah. These researchers found that the competitive ability of carboxylic acids in preventing phosphate sorption on kaolinite, gibbsite, goethite and different soils was in the order citrate > oxalate > malonate > tartrate > acetate. Researchers have demonstrated that the competitive sorption of phosphate and organic ligands (oxalate, tartrate, succinate and malate) on variable charge minerals was influenced by pH, nature of ligand, order of anion addition, and initial ligand/phosphate molar ratio. Maximum reduction in phosphate addition occurred when an organic ligand was introduced before phosphate, whereas minimum reduction occurred when phosphate was introduced before an organic ligand. The efficiency of organic ligands in preventing phosphate sorption was affected by the nature of the surfaces of soil components and soils.

Many studies have demonstrated that LMMOLs inhibit sulfate sorption much more than phosphate. Martinez *et al.* studied the effect of 13 LMMOLs (eight aliphatic and five aromatic) on the sorption of sulfate on two variable charge soils. These soils, containing large amounts of citrate-dithionite-bicarbonate (CDB)-extractable Al and Fe, sorbed significant amounts of sulfate at pH 4.0 and 5.0. All the aliphatic acids, but cis-aconitic acid, prevented the sorption of sulfate, at both pH 4.0 and 5.0, with oxalic, tartaric, and citric acid, showing the greatest effect. Conversely, the aromatic acids and cis-aconitic acid usually increased sulfate sorption under certain conditions, probably because once sorbed these acids promoted the protonation of surface hydroxyl groups, thus creating a more positively charged surfaces.

Competition in sorption among phosphate or sulfate with LMMOLs in ternary systems, received some attention. He *et al.* found that the sorption capacity of phosphate onto Al(OH) x-montmorillonite complexes was much greater than that of oxalate and tartrate. Adding equimolar amounts of oxalate and tartrate resulted in an oxalate sorbed/tartrate sorbed molar ratio of ~1. Anyway, in the presence of phosphate the oxalate/tartrate molar ratios were <1 and decreased with increasing phosphate concentration, indicating that tartrate competed with phosphate more than oxalate for the sites common to all the ligands. Further, tartrate and oxalate added as a mixture were much more effective in inhibiting phosphate sorption than was tartrate or oxalate alone, probably because more sites with high affinity for the organic ligands were occupied by tartrate and oxalate than by tartrate alone.

The mutual interaction of sulfate, oxalate, citrate, and phosphate in binary and ternary systems on synthetic and natural allophanic samples at different pH values (4.0–7.0) was studied by Jara. The sorption capacity of sulfate, citrate, and oxalate (in the order listed) was much lower than that of phosphate. The sorption of sulfate was strongly inhibited in the presence of equimolar amounts of oxalate, citrate, or phosphate even at low surface coverage of all the ligands. The lower the final pH of the binary systems, the greater the inhibition of sulfate sorption. On the contrary, sulfate very poorly prevented the sorption of the other ligands (usually lower than 10%). Figure shows that more citrate than sulfate was adsorbed at comparable pH values. Citrate strongly inhibited sulfate sorption mainly at pH < 6.0, whereas sulfate very poorly inhibited citrate sorption. In ternary systems (sulfate+citrate+phosphate) the sorption of sulfate decreased much more than that of citrate.

It has been demonstrated that root exudates strongly inhibit the sorption of toxic anions (selenite, chromate, arsenate and arsenite) onto different sorbents. Zhu *et al.* studied the effect of organic ligands on the sorption of arsenite and arsenate on ferrihydrite and noncrystalline Al precipitation product. The efficiency of the anions studied in preventing arsenite and arsenate sorption was as follows: Oxalate < malate ≈ tartrate < citrate.

The efficiency of most of the competing anions in inhibiting the sorption of the metalloid on ferrihydrite decreased by increasing the pH. In acidic environments, all the ligands prevented the sorption of arsenite more than that of arsenate, but in alkaline systems, arsenite and arsenate seemed to be retained with the same strength. In contrast, on amorphous Al oxides higher sorption of arsenate versus arsenite was always evidenced in a wide range of pH, both in the absence or presence of organic ligands. Caporale *et al.* evaluated the arsenite and arsenate sorption capacity of a Fe-Mg layered double hydroxide at varying pH values in presence of citrate, oxalate or tartrate. They observed that this sorbent was able to held a much higher amount of arsenate than arsenite and noted that the sorption of the former was much more pH-dependent than the latter. These researchers also found that greater percentages of arsenite than arsenate were replaced by selected LMMOLs.

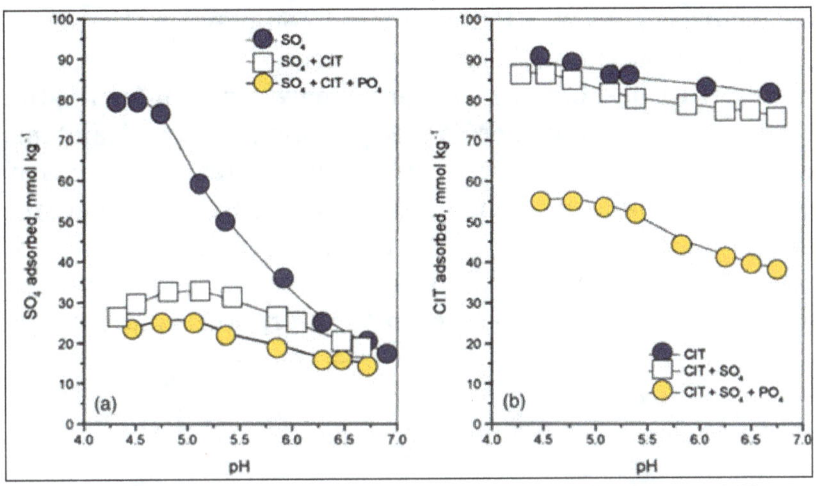

Sorption of (A) sulfate (SO_4) and (B) citrate (CIT) on a synthetic allophane at different pH values, when added alone or in a binary (SO_4+ CIT) or ternary (SO_4+ CIT + PO_4) systems.

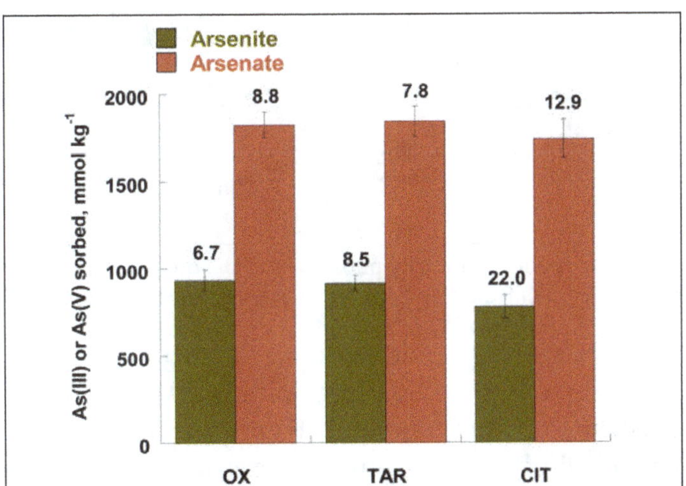

Arsenite and arsenate sorbed on Mg-Fe-LDH, in presence of selected LMMOLs [oxalate (OX), tartrate (TAR) and citrate (CIT)], at pH 6.0 and 20 °C, after 24 h of reaction time. As(III) or As(V) / LMMOL molar ratio of 1; As(III) and As(V) initially added = 1000 and 2000 mmol kg^{-1}, respectively. Number above bars indicate efficiency (%) of LMMOLs in inhibiting As(III) or As(V) adsorption by Mg-Fe-LDH.

Sorption/desorption of Nutrient and Pollutants on/from Organo-mineral Complexes

The reactivity of the colloids formed at soil-root interface results from complex interactions between clay minerals, (oxy)hydroxides, organic matter and/or microorganisms of soils. The sorption/desorption equilibria and kinetics of anionic and cationic species on/from natural colloids could substantially differ compared with their interaction with pure soil components.

The role of root exudates present in organomineral complexes on the sorption of anions and cations has received attention. In soil environments, depending on the mechanisms of their interfering reactions, organic substances have a dual role of both hindering or promoting anion retention by the Al or Fe precipitation products. Maintenance of short-range structure of the precipitates with a large surface area by the presence of critical concentrations of some biomolecules helps to promote a high retention capacity of organo-mineral complexes if compared with Fe and Al oxides formed in absence of organic ligands characterized by greater crystallinity and lower surface area, but the presence of organic ligands into the complexes also prevents the sorption of nutrients and pollutants.

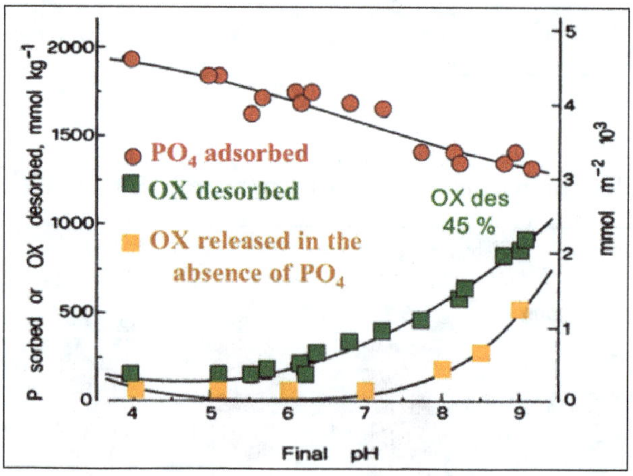

Effect of pH on the sorption of 2000 mmol/kg phosphate (PO_4) on an Al(OH)x-oxalate complex, and desorption of oxalate (OX) from the complex in the presence or absence of phosphate.

Violante and coworkers carried out some works on the sorption of phosphate and sulfate by organomineral complexes formed by coprecipitating Al in the presence of LMMOLs (oxalate, aspartate, citrate, tartrate, tannate). Figure shows the sorption of phosphate on and the partial removal of oxalate from a noncrystalline Al(OH)x-oxalate complex, containing 1625 mmol oxalate kg^{-1}, as a function of pH (4.0-9.0). Oxalate was only partially removed in the absence of phosphate. Negligible amounts of oxalate were desorbed in the pH range 4.0-6.0, but greater quantities were removed at pH > 7.0, due to the presence of OH^- ions. Phosphate replaced larger quantities of oxalate ions from the complex particularly in alkaline environments (oxalate desorbed 45% at pH 9.0). The quantities of sulfate sorbed on the same complex were much lower than those of phosphate, clearly because sulfate forms *outer-sphere* complexes in mildly acidic and alkaline environments and was not able to replace oxalate anions present in the organo-mineral complex.

In order to study the role of different LMMOLs present in synthetic organo-mineral complexes on the sorption of phosphate short-range ordered Al-oxyhydroxides with comparable surface area were synthesized in the presence of critical concentrations of selected ligands; aspartate (L/Al molar ratio [R] = 0.2), citrate (R = 0.01), tartrate (R = 0.01), tannate (R = 0.01 and 0.02). From 100 to 160 mmol kg^{-1} of ligands were present in the poorly crystalline boehmites formed in the presence of citrate, tartrate, or tannate; a much greater amount of aspartate (437 mmol kg^{-1}) was found in the sample formed in the presence of this ligand. The sample formed in the presence of aspartate retained more phosphate than the samples obtained in the presence of the other ligands, even though a much larger amount of organic anions was present in the former than in the latter. Clearly, aspartate which shows a moderate affinity for Al, would poorly counteract phosphate sorption and would be easily replaced by phosphate. On the other hand, citrate, tartrate, and tannate, which have a high affinity for Al, would not easily be removed and inhibit phosphate sorption. The amount of phosphate adsorbed by the precipitates after oxidation treatment (NaOCl) strongly increased due to the removal of counteracting organic anions as well as disgregation of aggregates promoted by organic ligands. These findings demonstrate that phosphate sorption on organomineral complexes varied with the nature and amount of biomolecules present in the coprecipitates.

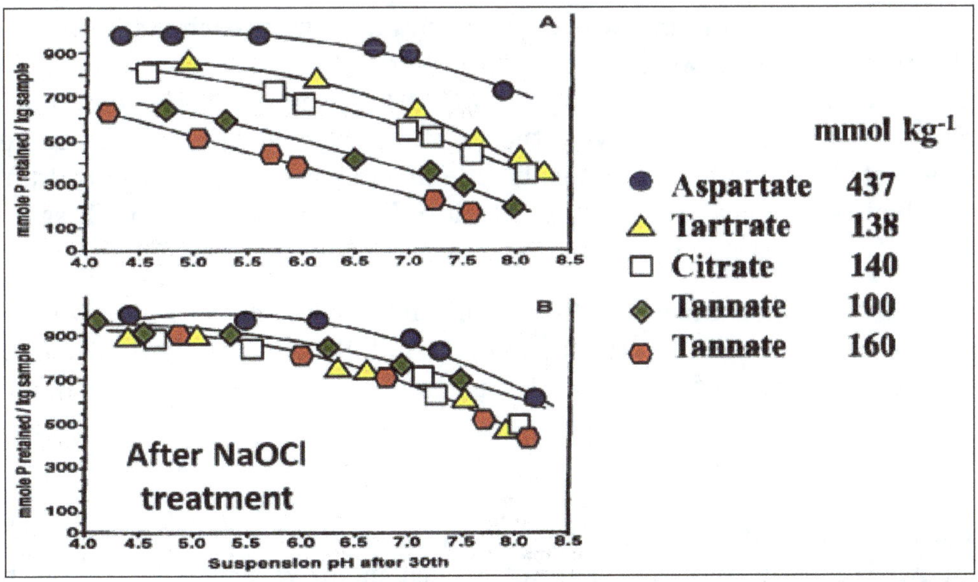

Sorption of phosphate by precipitation products of Al formed in the presence of organic ligands (A) before oxidation by NaOCl treatment and (B) after NaOCl treatment. In the figure A are reported the amount of each organic anion which promoted the formation of poorly crystalline Al-oxyhydroxides.

Martin *et al.* demonstrated that arsenite and arsenate equilibria and the kinetics of arsenite sorption on ferrihydrite-coated kaolinite as influenced by the presence of organic matter coverage. The hampering effect of the organic matter on arsenic sorption was attributable to the blocking by the organic ligand of the adsorbing sites, coupled with the steric hindrance of the large organic molecules, which restricted the arsenic to approach the surfaces.

More recently, Zhu *et al.* studied the sorption of cations [Pb, Cu and Cr(III)] added alone, in binary and ternary systems on an Fe-oxide (ferrihydrite) and on organomineral complexes obtained by coprecipitating Fe with equimolar amounts of oxalate and tartrate (tartrate+oxalate/Fe molar ratio of 0.1 or 0.2). Ferrihydrite sorbed much larger amounts of Cu, Cr and Pb thanthe organomineral complexes when mmol/kg mass scale was considered, owing to the larger surface area of ferrihydrite, whereas the opposite trend was true when mmol/m^2 surface scale was taken into account. Higher organic carbon content in the precipitates induced larger sorption (mmol/m^2) of heavy metals. Moreover, heavy metals were sorbed more strongly on organomineral complexes than that on ferrihydrite because the presence of organic ligands provided strong complexing sites and decreased the point of zero charge (PZC) of the sorbents.

The competition in sorption of the heavy metals was affected by the nature and the initial concentration of the elements, the surface properties of the sorbents, the addition sequence of the trace elements and the residence time. In binary systems the highest inhibition capacity on all the sorbents was for Pb versus Cu, whereas the lowest was for Cu versus Pb. The inhibition of Cr for Cu was much stronger than Cu for Cr. However, Cu competed with Cr and Pb more intensely on the organomineral complexes than on ferrihydrite, whereas Pb competed with Cr more intensely onto ferrihydrite. In ternary systems, when equimolar amounts of each metal were added, the surface of the sorbents were covered with more than 50% by Pb (from 53 to 56%), with 32 to 36% by Cr and with 9 to 13 % by Cu. However, even large amounts of Pb and/or Cr did not completely prevent the sorption of Cu, clearly because some sites on the sorbents had a high affinity for this metal. Only when the concentration of Cr was much higher than that of Pb, more sorption sites were occupied by Cr against Pb in binary and ternary systems.

The competitive sorption and desorption of Cu and Cr on/from goethite and goethite-B. thuringiensis complex was investigated by Zhu et al. Sorption isotherms and kinetics showed that the sorption of heavy metals was more and faster on goethite-B. thuringiensis complex than that on goethite and was suppressed by their competition. The competitive sorption between Cu and Cr was more intensive on goethite-B. thuringiensis complex than that on goethite and the competitive ability of Cr was stronger than that of Cu on both sorbents. Desorption observation of sorbed metals by $Ca(NO_3)_2$ showed that the desorption ratio of each metal was usually slightly higher from single system than that from binary systems. Lower desorption ratio of sorbed metal from goethite whereas relative higher desorption percentage from goethite-B.

thuringiensis complex confirmed higher affinity of heavy metals for the former sorbent than that for the latter one.

Effect of Root Exudates on the Sorption of Biopolymers

The possible effect of root exudates on the sorption of biopolymers have received some attention. Huang *et al.* demonstrated that citrate drastically inhibited the sorption of phosphatase on goethite, kaolinite and clay fractions of Chinese soils, clearly because citrate anions were able to form strong *inner-sphere* complexes on the surfaces of the sorbents and competed with the enzymatic molecules for sorption sites. With an increase in citrate concentrations sorption of enzyme decreased steadily. Conversely, acetate, which is weakly adsorbed on the surfaces of minerals and oxides, exerted no influence on enzyme fixation. These studies demonstrate that the presence of root exudates at the soil-plant interface may strongly influence the residual activity of enzymes by affecting their sorption onto soil components and organo-mineral complexes.

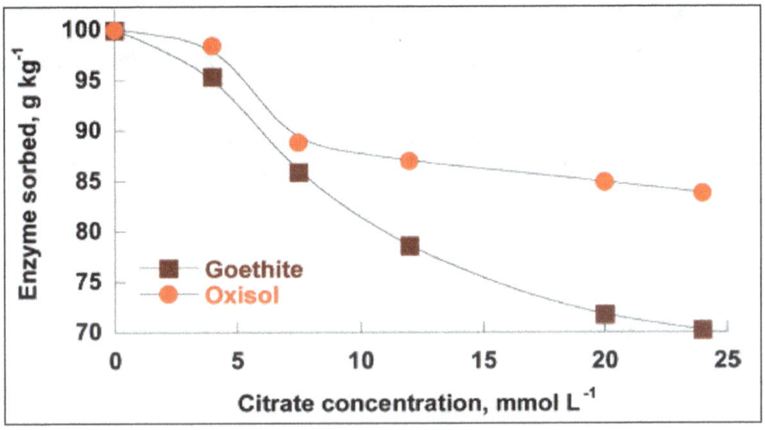

Amounts (%) of acid phosphatase sorbed on Oxisol clay and goethite (kept in suspension) in the presence of increasing concentrations of citrate.

Biomolecules produced by plants and microorganisms are involved in the weathering of minerals and subsequent release of Fe, Al, and other cations promoting the formation of Al and Fe precipitation products (iron plaques) and organomineral complexes. They perturb the hydrolytic reactions of Al and Fe and influence the crystallization, mineralogy, surface properties and reactivity of their final precipitation products. The coprecipitation of root exudates and Al and/or Fe facilitate the formation of short-range ordered or noncrystalline oxides of extremely small size (nanomaterials). Low molecular mass organic ligands and biopolymers (e.g. enzymes) interact with clay minerals, humic substances, OH-Al and OH-Fe species, and/or microorganisms, promoting the formation of organo-mineral complexes with peculiar surface properties, reactivity and activity (if enzymes are present). The close association of organic substances with short-range ordered Al- and Fe-oxides and phyllosilicates strongly stabilize organic C in soil environments including rhizosphere. Finally, root exudates influence the sorption and desorption of nutrients and toxic elements on/from abiotic and biotic soil components and organo-mineral complexes. They also affect the mobility and bioavailability of nutrients and pollutants and the toxicity of metals and metalloids.

References

- Bird, M. (2015). "Test procedures for biochar in soil". In Lehmann, J.; Joseph, S. (eds.). Biochar for Environmental Management (2 ed.). p. 679. ISBN 978-0-415-70415-1
- Soil-chemistry, earth-and-planetary-sciences: sciencedirect.com, Retrieved 21 March, 2020
- Soil-microorganisms, agricultural-and-biological-sciences: sciencedirect.com, Retrieved 15 August, 2020
- Batjes, Niels H. (1996). "Total Carbon and Nitrogen in the Soils of the World". European Journal of Soil Science. 47 (2): 151–63. doi:10.1111/j.1365-2389.1996.tb01386.x
- Ellis, Boyd; Foth, Henry (2017-03-09). Soil Fertility, Second Edition. pp. 73–74. ISBN 9781566702430. Retrieved 2017-05-19
- Soil-Microorganisms, The-Living-Soil, Building-Soils-for-Better-Crops-3rd-Edition, Learning-Center: sare.org, Retrieved 22 January, 2020
- Olness, A.; Archer, D. (2005). "Effect of organic carbon on available water in soil". Soil Science. 170 (2): 90–101. doi:10.1097/00010694-200502000-00002
- Blake, L. (2005), "ACID RAIN AND SOIL ACIDIFICATION", Encyclopedia of Soils in the Environment, Elsevier, pp. 1–11, doi:10.1016/b0-12-348530-4/00083-7, ISBN 9780123485304

4
Ecological Biochemistry

The chemical interaction of organisms with their abiotic environment is called ecological biochemistry. It further includes chemical ecology, pesticide biochemistry, pesticide chemistry, etc. This chapter discusses the subject of ecological biochemistry in detail.

Ecological biochemistry is concerned with the biochemistry of interactions between animals, plants and the environment, and includes such diverse subjects as plant adaptations to soil pollutants and the effects of plant toxins on herbivores. The intriguing dependence of the Monarch butterfly on its host plants is chosen as an example of plant-animal coevolution in action. The concept that chemistry and biochemistry may play significant roles in ecological relationships is not a new one. It is only within the last quarter of a century, however, that this concept has gained general acceptance and ecological biochemistry has become a well-defined interdisciplinary subject in its own right. Examples are given of different types of biochemical relationships involving plants and animals and reference to other areas of the subject are made.

Biochemical Indices in Ecology

Ecological diversity measurements are necessary to understand survival and adaptability of species. The role of biochemistry in interactions between organisms or between an organism and its environment (soil, water and air) cannot be overestimated.

Soil biological processes, such as organic matter degradation, mineralization, and nutrient recycling, are basically driven by microorganisms and enzymes. Community level physiological profiling using Biolog EcoPlates can be used as sensitive and effective indicator for evaluation of microbial functional diversity and community of soil under sewage sludge treatment. Also, the possibilty of early detection of soil physical and chemical properties from modification in biological properties, including microbial community has been documented by Szarlip, et al. The ratio of total Nitrogen calculated from biochemical soil properties (Nc) (biomass C, mineralised N, phosphomonoesterase, β – glucosidase and urease activities) to that measured by Kjeldahl method (Nk) was used as an index of soil quality in disturbed soils; in all cases, soil degradation was reflected in Nc/Nk values. The researchers recommended the adoption of the Nc: Nk for evaluation of soil degradation as it distinguishes among biochemically balanced, activity transient and degraded soils and its application in the rapid calculation of ecological dose, ED_{50} of soil pollutants. Soil productivity has been linked to biochemical activity. The quality of soil is related to the maintenance of equilibrum between organic matter content and these five biochemical indices. Nwaichi and Frac recorded correlation of the 5'-terminal restriction fragments generated after Csp digestion of 16S rRNA gene with observed DNA concentrations in the soil community profile and revealed loss of diversity with crude oil pollution. They posited that the herculean nature of the study of complex diversity pattern of the microbial community in parched and inundated soil environment may have arisen from inadequate biological classification of such environment. Molecular methods have the advantage of obtaining information about uncultivable organisms and can be applied to study complex trophic interactions in the field and to address underlying ecological questions. A wholistic understanding of soil microbial communities has been broadly defined by biochemical-based techniques and molecular-based techniques. Similarly, using Average well colour development in BIOLOG system with 95 different carbon sources used to produce a metabolic profile of microorganisms, Nwaichi and Frac showed wide distribution in soil community spread.

The ratio of RNA to DNA (known as standardized RNA/DNA ratio) has been widely used as a biochemical index to assess potential survival and growth and in determining the ecophysiological condition of marine organisms. Besides becoming a particularly promising biochemical tool that not only reflects physiological condition, this index also allows for the estimation of instantaneous growth rates in a great variety of organisms. Chícharo and Chícharo also reported validation of standardized RNA/DNA ratio for study of marine small animals. Cytochrome P450, a biomarker often used in the studies assessing aquatic environment pollution has proved to be a very suitable method for biochemical and environmental monitoring of aquatic environment quality. In Czech Republic, Marcela, et al. employed liver enzymes of phase I and phase II of xenobiotic transformation, namely cytochrome P450, ethoxyresorufin-O-deethylase, glutathione-Stransferase and tripeptide glutathione to assess the contamination status of different rivers. They reported that these biomarkers are very suitable and useful tool for monitoring of aquatic pollution in the study Czech rivers. The relationship between levels of biochemical markers in fish liver and the ambient level of organic pollutants in fish muscle and bottom sediment were demonstrated.

The use of Ascorbic acid content, chlorophyll content, leaf extract pH and relative water content have been used to determine the sensitivity and tolerance of plants to air pollution. In their study, Tanee, et al. concluded that air pollution tolerance indices of monitored plants were of the sensitive category and hence can be used to monitor air quality of Niger Delta. Based on their results, air pollution tolerance index, APTI obtained using the four leaf biochemical status can serve in predicting air quality condition of an area and these biochemical parameters are important in the study of plant-environment interactions as they can serve as bioindicators of air pollution. In the same vein, Krishnaveni and Magesh investigated air pollution tolerance index using some biochemical indices and submitted that most of study the plants were sensitive to parameters and advanced the added merits of these indices to include simplicity, cost-effectiveness of process and equipment. The plant species which are more sensitive act as biological indicators of air pollution. Increased number of important biochemical parameters gave more accurate and reliable results in air quality and exposure study caried out by Sarika, et al. Furthermore, their results from calculation of APTI values of twelve different tree species revealed that all monitored biochemical parameters are important and greatly govern in evaluation of APTI values and thus, selection of suitable tree species to remove/reduce air pollution.

Biochemical systematics and ecology has accelerated understanding of environmental assessment, monitoring and management and should be further explored in related field. Environmental biota, processes, function, and the interactions has been to a great extent, revolutionized by biochemical indices. In a time of great global environmental crises, increasing complexity of environmental problems and obvious strained resources of the regulators and polluters, most reliable approach to studying ecology has become an increasingly important topic.

Chemical Ecology

Chemical ecology examines the role of chemical interactions between living organisms and their environment, and the consequences of those interactions on the ethology and evolution of the organisms involved. It is thus a vast and highly interdisciplinary field. Chemical ecology studies focus on the biochemistry of ecology and the specific molecules or groups of molecules that function as signals to initiate, modulate, or terminate a variety of biological processes such as metabolism. Molecules that serve in such roles typically are readily diffusible organic substances of low molecular mass that derive from secondary metabolic pathways, but also include peptides. Chemical ecological processes mediated by semiochemicals may be intraspecific (occurring within a species) or interspecific (occurring between species).

The field relies on analytical and synthetic chemistry, protein chemistry, genetics, neurobiology, ecology, and evolution.

Chemical Ecology of Plants

Chemical ecology in plants is the study that integrates the chemistry and biological properties of the plants and their interaction with the environment (i.e. microorganisms, phytophagous insects) and their antagonists. Chemical ecology in plants mainly involves plants fighting herbivory by producing various phytochemical compounds.

These compounds can be produced by the plants themselves, or as a part of various symbiotic relationships fungi and bacteria.

The surface of the primary aerial parts of terrestrial plants is covered by a thin waxy structure known as the cuticle. The cuticle has crucial autecological functions and plays an important role as an interface in trophic interactions. The cuticle is composed of the cuticular layer and the cuticle proper, which is covered by epicuticular waxes. Whereas the cutin fraction is a polyester-type biopolymer composed of hydroxyl and hydroxy epoxy fatty acids, the cuticular waxes are a complex mixture of long-chain aliphatic and cyclic compounds. These highly lipophilic compounds determine the hydrophobic quality of the plant surface and together with the microstructure of the waxes, vary in a species-specific manner. The physiochemical characteristics contribute to certain optical features, limit transpiration and influence adhesion of particles and organisms, and as a result prevents it from undergoing wilting. Apart from that the cuticle acts like a skin for plants which prevents any mechanical damage to them from the external sources which either living organisms or any other abiotic component.

In many cases, the chemical ecology of plants involves mutualistic interactions with other organisms. One of these involves interactions with fungi, and in particular, mycorrhizae- where fungi form a sheath on the outside of the roots, or penetrate the roots growing between root cells, and even pushing through cell walls of individual root cells. In this relationship, fungi produce chemicals that decompose organic matter in the soil around the root, absorb the inorganic nutrients released by this decomposition thanks to much larger surface area of the fungi threads, compared to the absorbing surface of the root, and pass some of the water and nutrient the plant, thus greatly enhancing the ability of the plant roots to extract nutrients and water from the soil. The fungi may also provide a chemical protection (antibiotics) against harmful bacteria and fungi in the soil.

Interactions with Microorganisms

Plants interact also with micro-organisms. For this to become possible the microbes have to establish an interface between them and the plant, by growing into the plant through its surface. To do so, the microbes need to break the protective hydrophobic waxy layer on the plant's surface. To do this, the micro-organisms secrete special fluids which break down the fats from the cuticle.

Growth in Plants

Most of the hormones in plants are concentrated on their tips. The auxin hormones are responsible for growth of plants and are stimulated by certain stimulus such as light. This phenomenon is called phototropism, which is the movement towards or away from a light source. This growth enables the plant to obtain essentials such as sunlight which is very necessary for the photosynthesis. Therefore, the cuticle is one of the fundamental parts of the plant due to its physical and chemical properties such as waxy and thin like structure that enables it to be adapted to various mechanism such as hydrophobicity, interactions with microorganisms and growth of plants.

Plant-insect Interaction

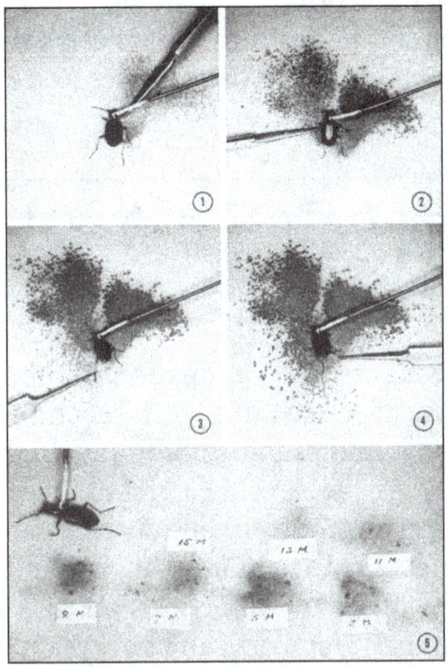

Series from a study by Eisner and colleagues investigating defensive spray in bombardier beetles. The paper is specially treated to have a color reaction with the spray, which is normally clear.

Monarch butterfly caterpillar on milkweed plant.

The chemical ecology of plant-insect interaction is a significant subfield of chemical ecology. In particular, plants and insects are often involved in a chemical evolutionary arms race. As plants develop chemical defences to herbivory, insects which feed on them evolve immunity to these poisons, and in some cases, re-purpose these poisons for their own chemical defence against predators. One of the more well-known examples of this is the monarch butterfly, the caterpillars of which feed on the milkweed plant.

Milkweeds contain cardenolide toxins, but monarch butterfly caterpillars have evolved to remain unaffected by the toxin. Instead, they sequester the toxins during their larval stage and the poison remains in the adult, making it unpalatable to predators. Many other such examples of this exist, including *Manduca sexta* (tobacco hawk moth) caterpillars which actively sequester nicotine found in the tobacco plant; and the bella moth, which secrets a quinone-containing froth from its head when disturbed by a potential predator obtained from feeding on *Crotalaria* species as a caterpillar.

Ecological Potential of Phytoremediation and Ecorestoration

Fly ash generates as the result of coal combustion in thermoelectric power stations whereas ore mining activities produce mine waste-rock and tailings worldwide. High concentrations of metal(loid)s and organic pollutants in fly ash and mine wastes are released into soil, air, and water presenting a global threat to the surrounding environment and human health. The environmentally sound management of fly ash and mine waste–rock and tailings includes monitoring stability of the dam construction and seepage flowrate, prevention of water erosion and dust spreading, reducing the footprint of the management facilities and successful restoration/revegetation. Harsh conditions prevailing on fly ash and mine deposits are unfavorable mechanical composition and pH, high concentrations of soluble salts, lack of nitrogen and phosphorous, reduced number of microorganisms and fungus, toxic concentrations of As, Au, Ag, B, Cu, Cd, Cr, Hg, Mn, Mo, Ni, Pb, Zn, and the presence of PAHs and PCBs.

Thermal power plants using coal as major fuel for production of electricity worldwide generate large amounts of combustion coal residues (CCRs), such as fly ash, bottom ash, boiler slag and flues gas desulphurization materials. Coal consumption in thermal power stations in world electricity supply should increase from 29.9% in 2011 to 46% in 2030. The coal reserves in the world are over 850 Gt with large producing countries: North America (240 Mt), Russian Federation (170 Mt), China (130 Mt), Europe and Eurasia (130 Mt), Australia (80 Mt), India (70 Mt), Middle East and Africa (40 Mt), Central and South America (30 Mt) and other Asian Pacific countries (10 Mt). The annual production of fly ash reaches approximately 580 million tons in China, 160 million tons and 130 million tons in US. Improper disposal of coal fly ash presents environmental and human health hazards. However, efforts have been made to develop different recycling technologies addressing fly ash utilization. Thus, fly ash can be used as building material, backfills for road constructions and pavement base, for cement and brick manufacturing, filler in asphalt mixture, soil stabilization as cover suitable for soil reclamation, material recovery, manufacturing mineral wool, and zeolite synthesis. Effective utilization of CCRs is 53% of total production.

The opencast coal and metal mining result in the formation of substantial amounts of different types of mine waste materials which depend on the geology at the mine area, physico-chemical composition, process technology, and mining operation (extraction, beneficiation, processing minerals) as well as how they are managed at the site. Some mine wastes include: (a) overburden that presents natural soil or massive rock which is removed to gain access to the ore deposits at open pit mines and it is settled on the surface at the mine site; (b) waste rock which contains industrial minerals which cannot be extracted as market products and they are usually stored in heaps and dumps on the mine site; (c) tailings: coarse (dry) fractions of mineral waste products which remain after processing material extraction for profit; fine (wet) fractions resulting from flotation as mineral separation technique used for getting valuable products from the coarse fraction. Coarse and fine fractions of mineral waste are settled into tailings pond and heap and enclosed by dams. However, sometimes, waste materials are turned into resources: overburden may be used for revegetation during mine closure, waste rock can be reprocessed to extract minerals and metals and used as backfill, aggregate in road construction and concrete whereas manganese and clay—rich tailings have been used in agro-forestry and construction materials (bricks and cement.

Coal fly ash is usually disposed by the "wet" method on the artificially made lagoons where fly ash and slag are mixed with water. From this suspension overflow and drainage wastewaters are generated and discharged indirectly through the drainage channels into the river. The drainage well-system is often located along the periphery of the embankment to prevent disturbance of the surrounding terrain and chemical pollution of groundwater. The raising of the peripheral embankments is achieved by hydrocycling of the ash, whereby the ash fractions are separated from the pulp and directed into a special device for "building" the dam. On the landfill, measures of protection are implemented in order to reduce the negative impact of fly ash on the environment. That is achieved by maintaining the drainage channels around the landfill, drainage wells throughout the entire landfill, maintaining the maximum surface of the water mirror (lakes) on the active cassette (25–55% of the surface), using a cannon system for watering dry surfaces inside the flat part of the landfill, using a spray system for watering embankments and grass cover and conducting biological recultivation i.e., sowing grass and planting woody and bushy plant species. The fly ash disposal is done usually in cassettes where one is active and others are passive, and can be used for the technical consolidation of fly ash and drainage, but also in the case of accidents or the cessation of fly ash.

The facilities of mine tailings and waste rock can be swimming-pool tailings ponds and small or large ponds over 100 or 1,000 hectares. Waste rock and tailings heap can be over 200 m high. Managing mine waste includes: discarding slurried tailings into ponds, surface water and groundwater; disposal of dry tailings or waste rock onto heaps or backfilling into underground mines or open pits; using them for the construction of tailings dam or land restoration. The choice of applied methods depends on cost, environmental impacts, and risk of failure. 'Best available techniques' (BAT) for the environmentally sound management of tailings and waste rock include management strategies, monitoring, and risk assessment such as life-cycle management (design, construction, operational phases, closure, and after-care phase), prevention and control of water erosion and dust spreading, acid rock drainage (ARD) management (active/passive treatment, addition of buffering minerals), drainage of ponds, seepage management, accident prevention (dam design and construction, raising and operation of pond, and monitoring stability of tailings pond), restoration/revegetation heaps or dams, reduction of footprint (backfilling of waste materials), mitigation of accidents, closure, and after-care of facilities. Therefore, BAT applies risk/safety and environmental management system (EMS).

Mining activities and coal production from surface mines as well as ore and fly ash disposal produce large changes in landscape, lead to land degradation, deterioration of air and water quality, near-total loss of vegetation, affecting negatively fauna, and human beings. Fly ash and mine waste-rock and tailings contain many metal(loid)s and persistent organic pollutants (POP)s which can be released in the environment leading to water, air and soil pollution. Furthermore, fine particles can be dispersed due to wind in the surrounding environment affecting human health i.e., they provoke irritation in eyes, skin, noise, throat and respiratory system leading to anemia, asthma, bronchitis and cancer diseases.

Limiting Factors for Plant Growth on Fly Ash and Mine Waste Deposits

Limiting factors for plant growth on fly ash deposits and mine spoil are unforable mechanical composition and pH, high concentrations of soluble salts, lack of essential nutrients, such as N and P, toxic concentrations of As, B, Cd, Cr, Cu, Hg, Mn, Mo, Ni, Pb, Se, reduced number of microorganisms that can fix nitrogen and reduced presence of mycorrhizae. Natural soil is different from fly ash and mine spoil mostly due to no or poorly developed soil structure throughout the profile, coarse texture with more sand fractions and less clay content whereas the content of N, P and other essential nutrients is very low.

Generally, unweathering fly ash showed sandy structure with small amounts of colloidal particles and low percentage of hydroscopic water that result in great water permeability and weak aggregation indicating adverse physical conditions for plant growth. Furthermore, high values of electrical conductivity (EC) indicate a large amount of soluble salts whereas pH values can vary from 4.5 to 12 depending on coal type. Thus, lignite and bituminous coal have alkaline fly ash, more Ca and less S. In addition, pH values from 7 to 10 enhance the release of As, Cr, Mo, Se, Sb, V, and W whereas Cd, Cr, Cu, Fe, Mn, Ni, Pb, Si, and Zn have maximum leachability pH under 7. The carbon content in unweathering fly ash is high probably due to incomplete combustion of coal and may partly compensate the lack of organic matter. The nitrogen content and available K_2O and P_2O_5 content in fly ash are low. However, total metal(loid)s concentration in fly ash, such as As, B, Cd, Cr, Co, Cu, Mn, Mo, Ni, Pb, Se can be toxic and can vary depending on coal, combustion conditions, operating parameters, transport system, leaching processes, effects on plants, and climatic conditions. Over time, proportion of total sand in fly ash is reduced and the clay fraction is increased whereas the percentage of hygroscopic water is increased and this all together provides good aggregation and capillarity. In weathering fly ash, EC values and pH decreased, content of organic carbon and total nitrogen. The content of some elements (As, B, Cu, Mn, Mo, Zn) in weathering fly ash decreased as a result of leaching and strong element binding by Al, Si, and Fe oxides, clay and organic matter. Taken together, plant species growing on weathering fly ash deposits improve physico-chemical properties of fly ash providing favorable conditions for further plant growth.

Mine wastes show great heterogeneity in their texture and physico-chemical properties in metallic mining area. The wide range of mine spoil characteristics is related to differences in mineralogical substrate and composition of mine residue which can come from different ore minerals and veins as well as from the exploitation procedures, plant cover and climatic conditions. Thus, on the more or less revegetated mining tailings the fractions of sand, silt and clay vary greatly. In addition, the mining wastes show different pH values from acid, neutral to slightly alkaline to moderately alkaline values. Furthermore, the range of EC is also wide where these values increased from the border of mine tailings to the plateau of barren area. The values of organic carbon and total nitrogen also vary as well as available content of K_2O and P_2O_5. The mine spoils on the tailings showed multi-element pollution with high total concentrations of As, Cd, Cu, Fe, Hg, Mn, Ni, Pb, Sb, Zn that exceed local background values and the maximum allowed values proposed by any country legalization as well as agricultural/residential/commercial/industrial use and they largely depend on geochemical partitioning.

Ecorestoration of Fly Ash and Mine Waste Deposits

Development of vegetation cover on fly ash and mine waste deposits is significant because plant species stabilize fly ash and mine spoil, prevent wind erosion, decrease mobility, toxicity, and dispersion of chemical elements in the surrounding environment, provide the organic substance which can bind contaminants and reduce transfer of pollutants in the food web. Therefore, ecorestoration of degraded sites is the process of renewing ecosystem stability and resilience after stress or disturbance with respect to its health, integrity and sustainability presenting a key issue in environmental science and ecological engineering.

Native plant species grow in a particular area over a long period of time without human intervention and possess certain characteristics that make them the best adapted to local conditions providing practical and ecologically valuable alternative for landscaping and ecorestoration projects. Phytomanagement of degraded sites encourages selection of native species that in the long run form self-sustaining plant communities that do not require much maintenance. Establishing successful vegetation cover from native plant species on disturbed land promotes natural revegetation. Thus, selected native plants should have the following characteristics: (a) Easy to establish, (b) fast growth, (c) deep root system, (d) capacity for nitrogen fixation, (e) tolerance to local climate; (f) tolerance to adverse physico-chemical conditions; (g) favorable growth with other species of the mixture; (h) capability to improve substrate fertility through organic matter and nutrient cycling.

Successful ecorestoration is achieved by establishment of the self–sustaining vegetation cover through human intervention by seeding the mixture of grasses and legumes, and planting shrubs and trees. In a temperate climate, grass and legume species that are the best for the revegetation of degraded sites belong to the genus of Festuca, Lolium, Agropyron, Poa, Medicago, and Vicia whereas the trees that provide good phytostabilization of substrates are Poplar, Robinia, Salix, Alnus, Betula, and Acer. However, in tropical region different plant genus are the best choice for the recultivation of degraded sites: Chloris, Pennisetum, Vetiveria, Paspalum, Cynodon (grasses); Leucaena, Acacia, and Sesbania (legumes) and Prosopis, Eucaliptusand Ricinus (trees). Legumes present keystone species because they provide nitrogen as the critical, most important resource on fly ash and mine waste deposits i.e., they perform nitrogen fixation in the atmosphere by using bacteria. Other plants such as Secale cereale, Avena sativa, Hordeum sativum, Triticum aestivum, Dactylis glomerata, Festuca rubra, Poa pratensis, Lolium multiflorum, Lolium perenne, Arrhenatherum elatius, Phleum pretense are important as nursery culture. These species act as pioneers, because they begin the process of revegetation providing erosion control, improve the physico-chemical composition of fly ash and mine spoil, retain moisture, and nutritional substances that will be later used by spontaneous colonizers.

Over time, plant species that are sown/planted or they spontaneously colonize fly ash and mine waste deposits bind substrate with a fibrous root system, and they spread by seeds and rhizomes forming a dense vegetation cover. In the initial phase of colonization, annual, and biennial plant species are dominant on the disposal. They are characterized by *r*-selection and with them the process of vegetation regeneration begins. According to Grime biennials can be competitive ruderals (C-R) and stress-tolerant ruderals (S-R). Annual plants have a short life span, rapid growth during the season and large production of seeds (Erigeron, Chenopodium) whereas biennials have a large amount of seeds, life cycle of several years, but they flower and fruit only once, after what they die (Oenothera biennis, Daucus carota, Verbascum phlomoides, Tragopogon dubius). Perennial, herbaceous plants spread and grow from the seeds, rhizomes, stolons, bulbs, and tubers. The rhizomes are used for the accumulation of reserve substances and survival during adverse periods, and serve as a link between the above-ground shoots (Calamagrostis epigejos, Phragmites communis, Festuca rubra, Tussilago farfara, Cirsium arvense, Heliatnhus tuberosus, Epilobium). Perennials can be stress-tolerant ruderals (S-R), stress-tolerant competitors (C-S), competitive ruderals (C-R) and “C-S-R”. Trees and shrubs that can grow on fly ash or mine waste deposits are Acacia, Acer, Azadirachta, Albizia, Amorpha, Cassia, Dalbergia, Eucaliptus, Fraxinus, Grevillea, Leucaena, Melia azedarach, Morus, Platanus, Pongamia pinnata, Populus, Phyllanthus emblica, Rosa, Rubus, Salix, Tamarix, Tectona grandis. These plant species are characterized by K-selection because they have long-life cycle and slow growth with high assimilates investment in vegetative growth. Trees and shrubs can be competitors (C), stress-tolerant competitors (C-S), and stress tolerators (S).

Vegetation Surveys on Fly Ash and Mine Waste Deposits

Floristic composition of plant species on fly ash and mine waste deposits provides insight into the environmental and floristic potential of these sites in the process of biological recultivation and primary and secondary succession. Vegetation surveys showed that a large number of plant species that are sown or spontaneously colonized these degraded sites belong to the families Fabaceae, Poaceae, Asteraceae, Chenopodiacea, Brassicaceae, Rosaceae.

Fly ash deposits at the thermoelectric plant station "Nikola Tesla-A" (TENT-A) in Obrenovac (Serbia): Active cassette of fly ash (A–D); passive cassette of fly ash 3 years after biorecultivation with sown legumes, grassses, and shrubs (E–I); passive cassette of fly ash 11 years after biorecultivation with spontaneous colonizers (J–N).

The plant species that are tolerant to adverse conditions on fly ash deposits in Europe are Achillea millefolium, Agrostis capillaries, Calamagrostis epigejos, Chenopodium rubrum, Cirsium arvense, Crepis setosa, Cynodon dactylon, Dactylis glomerata, Daucus carota, Echium vulgare, Erigeron canadensis, Eqiuisetum sp., Epilobium hirsutum, Festuca rubra, Hypericum perforatum, Linaria vulgaris, Lolium perenne, Lotus corniculatus, Medicago sativa, Melilotus officinalis, Oenothera biennis, Phragmites communis, Plantago lanceolata, Reseda lutea, Rumex acetosella, Polygonum lapathiholium, Pycreus glomeratus, Sonchus arvensis, Sysimbrium orientale Rumex obtusifolia, Salsola kali, Silene vulgaris, Sinapis arvensis, Solidago serotina, Senecio vulgaris, Sonchus oleraceus, Sorghum halepense, Trifolium repens, Tussilgo farfara, Vicia villosa, Verbascum phlomoides, Xantium strumarium, Acer pseudoplatanus, Amorpha fruticosa, Gleditschia triacanthos, Eleagnus angustifolia, Morus alba, Morus nigra, Robinia pseudoaccacia, Rosa canina, Rubus caesisus, Populus alba, Salix alba, and Tamarix tetandra. Furthermore, different plant herbaceous species colonize fly ash deposits: Amaranthus deflexus, Calotropis procera, Cannabis sativa, Cassia tora, Chenopodium album, Croton bonplandium, Cynodon dactylon, Eclipta alba, Limnanthe, Ipomea carnea, Parthenium hysterophorus, Saccharum bengalense, Sacharum munja, Saccharum spontaneum, Sida cordifolia, Solanum nigrum, Thelypteris dentate, Typha latifoli. In Hong Kong, plant species that can grow on fly ash deposits are Eleusine indica, Neyraudia reynaudiana, Tamarix chinensis, Chenopodium acuminatum, Fimbristylis polytrichoides, Pteridium aquiilinum, Panicum repens whereas, in Australia plants with high ecological potential to grow on fly ash deposits are Atriplex, Enchylaena tomentosa, Halosarcia, Mesembryanthemum, Nitraria billardieri and Scaevola colloris. Vegetation surveys on fly ash deposits in South Africa showed the following plant species: Amaranthus hybridus, Chamaecrista bienis, Chloris gayana, Cynodon dactylon, Cyperus esculentus, Digitaria eriantha, Eragrostis sp., Hyparrhenia hirta, Lepidium bonariensis, Lespedeza cunea, Brachiaria serrata, Heteropogon contortus, Tristachya leucothrix, Setaria sphacelata.

Mine waste in Serbia: Copper mine waste in metalliferous Bor region (A–C) with Epilobium dodonaei (D), Calamagrostis epigejos (E), and Robinia pseudoaccacia (F);

mine flotation tailings at the Pb/Cu/Zn mine "Rudnik" (G–H) with spontaneously colonized Equisetum sp. (I); antimony flotation tailings "Stolice" after accidental flooding event with spontaneously colonized plant species 1 year later (J–O).

Plant species that have high capacity to grow and survive in mining areas are Agrostis stolonifera, Calamagrostis epigejos, Cerastium arvense, Polygonum aviculare, and Tussilago farfara in flotation tailings from Cu ore in Poland Agrostis capillaries, Rumex acetosella, Vicia hirsute, Apera spica-venti, Chenopodium botrys, Xantium italicum, Equisetum palustre, Persicaria lapathifolia, Vulpia myuros, Polygonum lapathifolia, Betula pendula, Populus tremula, Populus alba, and Populus nigra in in Timok River floodplain that are partially damaged by slurry sulphidic waste from Cu mine in Bor in Serbia; Agrostis stolonifera, Epilobium dodonaei, Calamagrostis epigejos, and Centaurea arenaria on non-reclaimed overburden sites, Robinia pseudoacacia, Cirsium eriophorum, Festuca valeisaca, Linaria genistifolia, Taraxacum offinale, Convolvulus arvensis, Achillea millefolium, Vicia cracca, Daucus carota, Centaurea stoebe Poa pratensis, and Rumex crispus on reclaimed overburden in Cu mine in Bor in Serbia; Dittrichia viscosa, Cistus salviifolius and Euphorbia pithyusa in mining tailings in Sardinia; Digitalis purpurea, Mentha suavolens, and Ruscus ulmifolius in abandoned Pb mine in Portugal; Zygophyllum fabago, Helichrysum decumbens, Tamarix, Lygeum spartum, Piptatherum miliaceum, Pinus halepensis, Tetraclinis articulate in Cartagena-La Union mining district, Spain; Coincya monensis, Agrostis durieui, Holcus lanatus, Festuca rubra, Dactylis glomerata, Cytisus striatus, Genista legionensis, Lotus corniculatus in Pb-Zn and Hg-As mining waste in Spain; Pistacia, terebinhtus, Cistus creticus, Pinus brutia, and Bosea cypria in Cu containing mine tailings in Cyprus; Ricinus communis in mine tailings, Mexico; Amaranthus watsonii, Solanum lumholtyianum, Bromus catharticus, Acacia farnesiana, Gnaphalium leucocephalum, Brickellia coulteri, Baccharis sarothoides, Prosopis velutina, Boerhavia coulteri from abandoned mine tailings, Sonora, Mexico; Salsola collina, Festuca elata, Medicago sativa, Ipomea purpurea, Grewia biloba, Cotinus coggygria, Zizipus jujube, Vitex negundo, Bidens parvifloa, Sonchus oleraceus, Artemisia annua and Carex tristachya on soil-rock mixture in abandoned mines in Beijing in China; Digitaria sanguinalis, Erigeron canadensis, Phytolaca acinosa, Pteris multifida, Cynodon dactylon, and Melastoma dodecandrum in Mn mineland in China.

Phytoremediation

Phytoremediation presents an efficient, "green clean," environmental, low cost, and eco-friendly technology that uses plants to reduce or remove inorganic and organic pollutants from environment. The main phytoremediation technologies are phytostabilization and phytoextraction, rhizodegradation/phytostimulation and phytodegradation/phytotransformation.

Phytoremediation technologies: Phytostabilization; Phytoextraction; Phytostimulation/Rhizodegradation; Phytodegradation/Phytotransformation. MT, metal(loid)s;

OC, organic compounds. Pollutants can be stabilized or degraded in the rhizosphere, sequestered or degraded in plant roots and leaves (A). Strategies for As, Cu, Zn, Mn, Fe, Cd, Ni and Pb uptake by plant roots from soil through different active transporters (ATs); ZIP/IRT, zinc, iron regulated protein; COPT, copper putative transporter; NRAMP, natural resistance –associated macrophage protein; HMA-heavy metal P-type ATPase; H^+ATP, HA-proton efflux transporter; FRO-ferric reductase/oxidase; PEZ_1, phenolics efflux $zero_1$; TOM_1, efflux transporter; YS_1, yellow stripe-$like_1$; NA, nicotinamine; MA, family of mugineic acids; LCT_1, low-affinity cation transporter; GNGC channels, putative transition metal transporter (B,C); Tolerance mechanisms for metal(loid)s in plants: (1) binding of metal(loid)s to the cell wall, (2) chelate of metal-organic acid on the outside of the cell membrane, (3) active metal efflux from cells, (4) metal-chelator complexes (5) compartmentalization or sequestration in a vacuole; Ats, active transporters; MTs, metal(loids); OA, organic acid; NA, nicotinamine; AA, amino acids; GSH, glutathione; PC, phytochelatines.

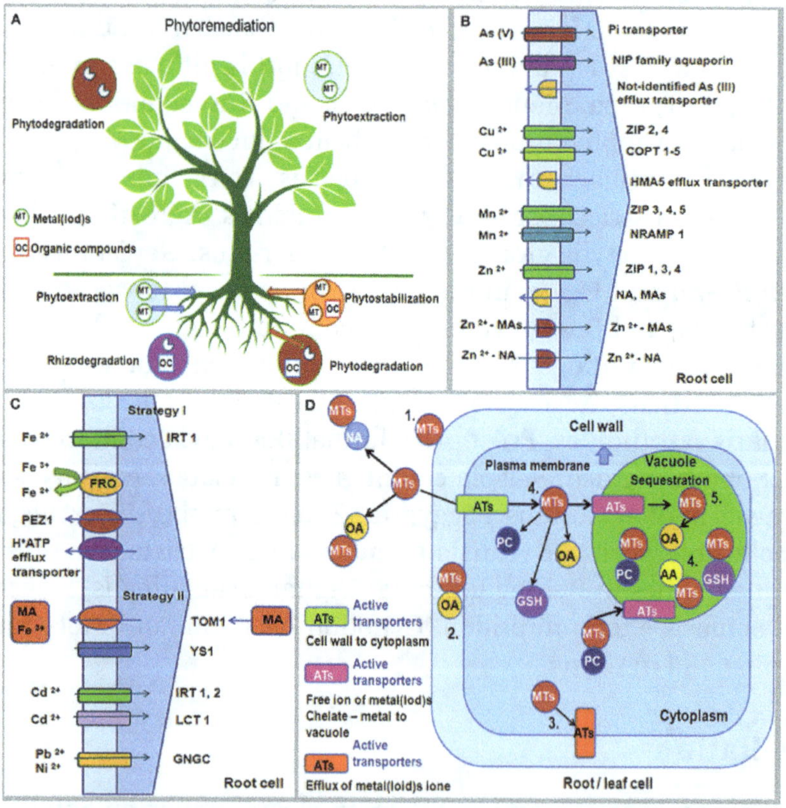

Phytostabilization is a phytoremediation technology which uses plants to reduce the mobility and bioavailability of metal(loid)s and organic pollutants in the surrounding environment preventing their migration in the soil and groundwater or their entry into the food web. In addition, phytostabilization improves the physico-chemical characteristics of the polluted substrate by increasing the content of organic matter and nutrient levels and preventing wind erosion. This can be achieved by using plant species that are named “excluders,” which are capable to limit the uptake of pollutants in high concentrations inside the plant tissue over a wide range of soil concentrations i.e., they immobilize pollutants in the substrate or in plant roots. The main mechanisms of phystostabilization are avoiding or excluding the pollutant from further transport at the root level by binding to the cell wall and/or preventing the transport through the plasma membrane, by the presence of mycorrhizal fungi and the presence of root exudates in the rhizosphere.

Phytoextraction is a phytoremediation technology that uses plants capable of accumulating metal(loid)s in the roots and aerial parts of the plants. This phytoremediation technique uses accumulators/hyperaccumulators because these plants survive regardless of excess concentration of metal(loid)s in their above-ground tissues. Therefore, plants suitable for phytoexctraction should have fast growth, high biomass, extended root system, high root-shoot transfer and good tolerance to high concentrations of metal(loid)s in the plant tissue. Generally, hyperaccumulators are not popular for phytoremediation due to slow growth rate and low biomass. Uptake of pollutants from the environment and their accumulation is achieved through the formation of chelates and sequestration of the metal(loid)s in the vacuole of roots and leaves.

Rhizodegradation/phytostimulation is phytoremediation technology that uses roots of plants which enhance microbial and fungal activity in the rhizosphere and breakdown organic pollutants (polycyclic aromatic hydrocarbons, PAHs and polychlorinated biphenyls, PCBs). According to Pilon-Smits and LeDuc suitable plant species should have large and dense root systems and rhizobacterial strains. Therefore, plants facilitate degradation of organic pollutants by stimulating microbial and fungal activity releasing exudes (organic acids, sugars, amino acids, phenolics, and enzymes dehalogenase, nitroreductase, peroxidase, laccase). Stimulation of soil microbial communities by root exudates provides acquisition of nutrients, changes of pH, water flux, availability of oxygen around roots and root growth. Plant growth promoting bacteria (PGPB) present a consortium of bacteria that colonize different niches of plant roots and they can degrade pollutants more efficiently than a single species/strain. Rhizobacteria, such as Achromobacter, Arthrobacter, Azotobacter, Azospirillum, Bacillus, Enterobacter, Pseudomonas, Serratia, and Streptomyces have been found to have beneficial effects on plant species in metal-contaminated environments because they can reduce the metal toxicity by biosorption and bioaccumulation.

Phytodegradation/phytotransformation is a phytoremediation technology which implies degradation of organic compounds by plant enzymes within roots or leaves to simple molecules, such as CO_2 and H_2O. The network of detoxification, degradation and storage of organic compounds in plants can be divided into three steps: (a) transformation (activation and transformation of organic compounds); (b) conjugation process which occurs between organic compounds and malonic acid, D-glucose, glutathione, and amino acids where conjugates become less mobile and less toxic; (ac) storage/elimination (sequestration or storage of organic compounds in cell vacuole, apoplast, or cell walls). This cascade of metabolic reactions is also named the "green liver model" due to its similarity to the mammalian metabolism.

Uptake, Transport, Tolerance and Cross—Talk between Metal(loid)s in Plants

The uptake of metal(loid)s depends on total and available concentrations, cation exchange capacity, redox potential, pH, the presence of Al, Fe and Si oxide, organic matter, texture, aeration and moisture of soil, and climatic conditions. Adaptive mechanisms of plants that are related to the uptake and translocation of pollutants can be different and they are associated with distribution, redistribution, storage, and detoxification of metal(loid)s in roots and leaves. Adaptive responses of plants to high or low concentrations of chemical elements include changes in the expression of the transporter genes responsible for the uptake, efflux, translocation, and sequestration of elements in plants. Transport of metal(loid) ions from roots to leaves is performed via membrane transporters, amino acids and/or organic acids. The maintenance of metal-ion homeostasis and plant tolerance in plants is enabled by some families of transporters in roots and leaves: a) Influx transporters: ZIP/IRT family of proteins (zinc-iron regulated protein permease) and NRAMP protein (natural resistance-associated macrophage protein); b) efflux transporters: HM—ATPase, CDF (cation diffusion facilitator), CAX (cation exchanger) and ABC (ATP-binding cassette transporters).

Plant tolerance to deficiency or toxicity of metal(loid)s can be achieved by excluding the element from further course of transport or its accumulation in plant cells. The exclusion mechanism is based on the limited absorption and accumulation of toxic metal(loid) ions through: (a) Increasing their binding capacity to the cell wall, (b) preventing their entry into the cytoplasm and transport through the plasma membrane by reducing the absorption through modification of the ion channels and/or ejection of the metal(loid)s from the cell by active efflux pump, as well as forming from the outer side of the plasma membrane metal—chelate complexes with organic acids (OA). Tolerance to high metal(loid) concentrations in cell cytoplasm (accumulation) is achieved through: (a) Inactivation by forming complexes between different chelate compounds (glutathione, GSH; phytohelatines, PC; organic acids, OA and amino acid, AA) and metal(loid)s; (b) compartmentalization in vacuoles and other cellular organelles (chloroplasts, mitochondria, Golgi apparatus, endoplasmic reticulum) and/or sequestration of metals by chaperones (proteins involved in intracellular transport of metals to the place of activation).

Therefore, As (V) is taken up through the phosphate transport system (As—Pi) whereas As (III) is adopted by the NIP family of aquaporin proteins (nodulin 26-like protein). As can be loaded into xylem and transported by the silicon transporter (Lsi2). In the cell, As forms a complex with glutathione (As—GSH) that can be transported in a vacuole through an unidentified ABC transporter or can form a complex with phytochelatines (As—PCs). In the condition of excess B, BOR2, and BOR4 transporters are accumulated in the plasma membrane of root cells. B can be exported from root cells into xylem reducing B concentration in roots. In roots and leaves, B is bound to the cell wall in the form of B-RGII complex which mitigates boron toxic effects on metabolism and plant growth. In addition, tonoplast intrinsic protein (TIP), such as AtTIP5;1 localized to the vacuolar membrane is more expressed under high B condition in leaf tissues.

Furthermore, Cu is taken up into the root cells through ZIP2 and ZIP4 transporters and putative copper influx transporters (COPT1), and then transferred into the cell via other transporters, such as COPT3 and COPT5 proteins, HM—ATPases and chaperones to the vacuole, chloroplast, mitochondria and Golgi complex, where it is sequenced: HMA5 protein is responsible for efflux Cu from root cells and its loading to xylem; HMA7/RAN1 (responsive-to-antagonist1) is a transporter in roots and leaves that has the role of delivering Cu to Golgi complex; HMA6/PAA1 transports Cu to the membrane of chloroplast, and further via chaperones (CCs) to Cu/Zn SOD; HMA8/PAA2 protein transports Cu through thylakoid membrane to plastocyanines (PC); COX chaperones deliver Cu into mitochondria; ATX1 delivers Cu to HMA5 protein, and CCH delivers Cu to Golgi complex.

High content of Mn is taken up by roots through ZRT/IRT1 transporters whereas in the condition of Mn deficiency ZIP3-ZIP7, NRAMP1 and OsNRAMP5 transporters are active. In the conditions of Mn deficiency, OsNRAMP5 transporter is also necessary in the accumulation of Mn in leaves whereas NRAMP3/NRAMP4 proteins transport Mn in the vacuole of the leaf and they are essential for the maintenance of photosynthetic efficiency. Intracellular transport and accumulation of Mn and its sequencing in root and leaf cells are achieved through proteins: P-type ATPase transport Mn into the endoplasmic reticulum (ECA1 – excess Mn, EC3 – Mn deficiency) and Golgi complex (PMR1); MCF family of proteins (mitochondrial carrier family) through chaperone MTM1 transport Mn to MnSOD; Metal/H^+ antiporter (CAX2), CDF family of transporters (ShMTP1), and ABC transporters transfer and accumulate Mn in the vacuole, thereby maintaining the homeostasis of Mn and providing its optimal amount for the metabolic processes in the plant cells.

Uptake, transport, intracellular sequencing and maintenance of homeostasis Zn in plant cells are regulated by numerous membrane transporters and chelating agents. In the conditions of Zn deficiency in Graminae, the secretion of phytosiderophores (mugenic acid families, MAs) is increased whereas in other plants the biosynthesis of nicotinamine (NA) is dominant. MAs and NA are excreted in the rhizosphere by specific transporters and have a high binding affinity for Zn. Formed Zn—MAs and Zn—NA complexes are further taken up and transported through the root cell membrane. In the conditions of Zn deficiency, numerous ZRT—IRT transporters are active (ZIP 1, 3 9–12 take up Zn from the soil; IRT3, ZIP4, 5, 8 transport Zn from root to leaf. Furthermore, P—type ATPases are responsible for Zn flow in apoplast and xylem, accumulation of Zn in plastids and its sequencing in the vacuole and under the condition of high zinc content, HMA1 transporter contributes to the detoxification of Zn in chloroplast;

CDF family of transporters (Metal Tolerance Proteins such as MTP1, 2, 3) transport Zn in the vacuole of leaves and roots); MFS family of transporters (Major Facilitator Superfamily of transporters), such as zinc induced facilitator (ZIF1) protein is localized in the tonoplast of the vacuole and is capable of transporting organic acid, Zn-organic acid complex, or NA in the vacuole; amino acid histidine and organic acids can form Zn complexes (Zn—histidine, Zn—malate and Zn—citrate) in cytoplasm and the vacuole of roots and leaves whereas PCR2 (cystein—rich protein) exports Zn from root to leaves.

Fe acquisition mechanisms are different in dicotyledonous (Strategy I) and graminaceous plants (Strategy II). In Strategy I, plants acidify the soil and reduce Fe^{3+} to soluble form Fe^{2+} that is transported across the root plasma membrane through ferric reductase oxidase 2 (FRO2) and 3 (FRO3) transporters. Fe uptake by roots also occurs through the IRT1 transporter. Excretion of proton (H^+-ATPase i.e., HA-proton efflux transporter) and phenolic compounds from roots to rhizosphere (phenolics efflux transporter, PEZ1, PHENOLICS EFFLUX ZERO 1) may increase the solubility of Fe ions or support the reducing capacity of Fe on the root surface. In Strategy II, graminaceous plant response includes secretion of MAs; efflux transporter TOM1 in the rhizosphere solubilizes Fe^{3+} and forms Fe^{3+}–MA complex that is taken up into root cells by the YELLOW STRIPE 1 (YS1) and YELLOW STRIPE 1–like (YSL) transporters. Fe translocation in plants includes chelators such as citrate, NA and MAs. However, Fe can be effluxed into the xylem by AtFPN1/AtIREG1 transporters localized in the plasma membrane. Influx transporters, such as YSL family members (YSL1, YSL2, YSL3, YSL15) are responsible for long-distance transport of NA-chelated iron into leaves. Furthermore, TOM1, IRT1, ENA1, ENA2 (nicotianamine 1 and 2) as efflux transporters are also important for Fe translocation inside the plant. According to Thomine et al. NRAMP3 and NRAMP4 transporters are significant in long distance metal transport. Fe mobilization in the chloroplast is enabled by FRO6,7 (ferric-chelate reductase oxidase) and MAR1/IREG3 (multiple antibiotic resistance 1/iron-regulated protein 3), that are both localized in envelope membranes; PIC1 protein (chloroplast permease 1) transports Fe across the inner envelope membrane of the chloroplast and regulates its homeostasis; vacuolar iron transporter 1 (VIT1) is proposed to efflux Fe from cytosol into vacuole.

Lead (Pb), chromium (Cr), and cadmium (Cd) do not have physiological functions in plants and can be toxic at low concentrations whereas nickel (Ni) is essential and cobalt (Co) is a beneficial microelement for plants. Major transporter system responsible for the uptake of Fe^{2+}, Mn^{2+}, Zn^{2+}, Ca^{2+} is ZIP/IRT, but according to Clemens the uptake of Cd from soil seems to occur via same transporters due to chemical similarity. Therefore, IRT1, IRT2 can be primarily responsible for Cd uptake into the root, accumulation and translocation. In addition, Cd^{2+}/H^+ and CAX2 antiporters might be involved in the accumulation of Cd in the vacuole of the root. The LCT1 (low-affinity cation transporter) is important for the uptake of Cd whereas GNGC channels are permeable for Ni and Pb. Plants in the condition of Pb excess and nutrient deficiency activate the same H^+/ATPase pumps. However, the main pathway by which Pb enters the roots is Ca^{2+}-permeable channel due to competition between two cations. Furthermore, NRAMP1,3 transporters are able to transport Cd and Co in roots and shoots. According to Hall and Williams HMA1-4 transporters are involved in the efflux and transport of $Co^{2+}/Cd^{2+}/Pb^{2+}$ whereas HMA9 is responsible for the export of Pb and Cd from the cell. ABC transporters (MRPs) are responsible for the Cd accumulation in the vacuole of the root in the form of chelates PC-Cd and GS-Cd; CDF transporters are involved in sequestration of Cd, Co and Ni in the vacuole of leaves; CBP4 protein shows tolerance to Ni and hypersensitivity to Pb and it is responsible for reduced Ni concentration and increased Pb content in plant cells. Some magnesium transporters, such as MGT1 show high affinity for Ni and its transport and accumulation in roots and leaves whereas DTX1 (protein for Detoxification 1) belongs to MATE family of transporters (multidrug and toxic compound extrusion) and it is capable of detoxifying Cd.

The determination of bioconcentration factor (BCF) and translocation factor (TF) is important for assessing the phytoremediation potential of plants as well as the mechanisms of tolerance and the survival strategy of plants on degraded sites. Bioconcentration factor (BCF) defines the relationship between the content of the chemical element in the root and its content in the substrate, providing information about element accumulation in the roots and removal efficiency of the chemical element from the given substrate. Translocation factor (TF) defines the relationship between the content of the chemical element in the leaves and its content in the root providing information about the efficiency of plants to transport metal(loid)s from root to leaves. It is stated that plants having BCF >1 and TF <1 show phytostabilization potential. In addition, plants with BCF <1 and TF higher than one as well as BCF and TF <1 also show phytostabilization potential. Plants that have all these combinations are excluders, because they possess mechanisms that maintain low uptake of soil-metal contents and small shoot-metal contents. However, plants with BCF and TF >1 are suitable for phytoexctraction and they are accumulators/hyperaccumulators.

Phytoremediation of Fly Ash Deposits

Plant species growing on fly ash deposits showed high phytostabilization or phytoextraction potential for different chemical elements. Thus, sown species Festuca rubra on fly ash deposits can be considered as an excluder suitable for phytostabilization due to limited transport of As, Cu, Mn, and Zn from root to leaves (BCF < 1, TF < 1) in spite of efficient transport of B in leaves (BCF < 1, TF > 1); Ricinus communis can be considered as a good phytostabilizator of fly ash for Cd, Cu, Ni, Pb, and Zn due to BCF < 1 and TF < 1; Saccharum munja is an excluder plant for Cd, Cr, Cu, Ni, Pb, Zn (BCF < 1, TF < 1), Fe and Mn (BCF < 1, TF > 1) indicated that this plant retained more elements in roots than in leaves or in spite of efficient transport of some elements from root to leaves, this species possesses mechanisms that prevent the uptake of high content of metals in plants; Typha latifolia is a good phytostabilizator for Cd in fly ash (BCF < 1, TF < 1) as well as for Cr, Cu, Ni, Pb, and Zn (BCF < 1, TF > 1) whereas for Mn this species can be considered as a good accumulator (BCF > 1, TF > 1); Ipomea carnea showed high phytostabilization potential for Cu (BCF < 1, TF < 1), Ni, Pb (BCF > 1, TF < 1), Cr (BCF

< 1, TF > 1) whereas for Cd (BCF > 1, TF > 1) this species can be suitable for phytoextraction; Thelypterys dentate is an excluder plant for Fe, Si (BCF < 1, TF < 1) as well as for As, Cd, Pb (BCF > 1, TF < 1); Sida cardifolia is suitable for phytostabilization because of larger content of Co, Ni (BCF < 1, TF < 1) and Pb (BCF > 1, TF < 1) in roots than in leaves whereas it is an accumulator for Cd, Cr, Cu, Fe, Zn (BCF > 1, TF > 1); Cynodon dactylon is also an excluder plant suitable for phytostabilization of Fe, Ni (BCF < 1, TF < 1), Pb (BCF >1, TF < 1), Cd, Cu, Mn (BCF < 1, TF > 1) because it maintains low uptake of metal content and relatively small shoot-metal content in plants; Chenopodium album possesses an excluder mechanism for Cd, Cr (BCF > 1, TF < 1) and Mn (BCF < 1, TF > 1) which makes it a good candidate for phytostabilization of fly ash whereas for Co, Cu, Fe, Ni, Pb, and Zn (BCF > 1, TF > 1) this species can be marked as an accumulator suitable for phytoextraction.

Plant species that can grow on fly ash and possess a high rhizomediation potential for organic compounds are Agropyron, Cynodon dactylon, Festuca arundacea, Festuca rubra, Lolium perenne, Lotus corniculatus, Trifolium pretense, Melilotus officinalis, Hellianthus annus. Plant promoting bacteria (PGPB) is characteristic for F. arundacea whereas Rhizobium leguminosarum was found in L. perenne and T. pretense. Bacterial strains that enhance the degradation of PAHs for Sorgum bicolor are Pseudomonas montellili, Bacillus subtilis, B. pumilis, Pseudomonas pseudoalcaligenes and Brevibacterium halotolerans. In addition, Salix inoculated with Pseudomonas sp., Janthinobacterium lividum, Serratia marcescens, Flavobacterium sp., Streptomyces sp., and Agromyces sp. showed high potential for rhizodegradation of PAHs. Furthermore, Medicago sativa and Lolium perenne show high potential in phytoremediation of PCB compounds because they enhance the number of microbial populations inoculated by Glomus sp. (arbiscular mycorrhizas, AMF).

Plants that can grow on fly ash deposits and have high potential for phytodegradation of PAHs are Daucus carota, Festuca arundacea, Lolium multiflorum, Amaranthus sp., Raphanus sativus, Ipomea sp., Sorghum drummondii, and Chenopodium rubrum(benzo(a)pyrene. In addition, plant species that are capable to metabolize the PCBs to the bound residue of cell walls by enzymes UDP-dependent glycosyltransferases (glucose, glucuronic acid) or by peroxidases and polyphenol oxidases (lignin, pectin, hemicelluloses) are Trifolium repens, Triticum aestivum, Hordeum vulgare, Daucus carrota, Atriplex sp., Rosa sp.

Phytoremediation of Mine Waste Deposits

Many plant species growing on mine waste rock and tailings showed great potential for phytoremediation. Thus, Cistus ladanifer is suitable for phytostabilization of mining areas and can be considered as an excluder plant for Al, Ag, Ba, Bi, Sr, and Sb; Ricinus communis is an excluder for Cd, Cu, Mn, Pb, and Zn in spite of high concentrations of Mn in shoots; Coincya monensis is considered as Zn hyperaccumulator whereas, Holcus lanatus, Festuca rubra, Dactylis glomerata present good excluders of Hg and As on the Pb-Zn and Hg-As mining area; Polygonum aviculare grown at slightly acid spoil on mine tailings with Ag, Au, and Zn accumulated Zn is near the criteria for hyperaccumulators; Epilobium dodonaei for Cu, Pb, and Zn, Helichrisum decumbens, Zygophylum fabago and Lygeum spartum for Pb, Cu, and Zn, Acacia retinoides, Helianthemum syriacum, Pinus halepensis, Teucrium capitatum, and Thymelea hirsuta for As, Cd, Pb, and Zn and Euphorbia pithysa for Pb, Cu, and Zn showed low element concentrations in shoots despite highly contaminated mine spoils and they are suitable for phytostabilization; Cynodon dactylon and Sorghum halepense are good for phytostabilization of Pb-Zn mine waste; Ditrichia viscosa can be a phytoextractor for As and Zn; Acer pseudoplatanus, Fraxinus excelsior, Robinia pseudoacacia, and Salix sp. are suitable for phytostabilization of mine waste.

Microbial and fungal activity is important for restoration of mine tailings, establishment of plants and biogeochemical cycling, particularly in acid mine drainage. Microorganisms change the bioavailability of heavy metals, solubilization of minerals in mine tailings, release nutrients, supply them to the plants, and enhance plant growth. Plant-growth promoting bacteria (PGPB), such as Bacillus arsenicoselenatis, Chryssiogenes arsenates and Sulforospirillum arsenophyllum have a high potential for mobilizing As in mine tailings i.e., they are capable to take up As from substrate and lower its redox state. Mummey et al. showed significant enrichment of bacterial, fungal, and total microbial biomass after 20 years of reclamation on surface mines. Different strains of plant growth promoting rhizobacteria were isolated from mine spoils in Lotus sp., Elymus repens, Bromus tectorum, and Cardaria draba as well as in Sedum alfredii and Medicago sativa. Mycorrhizal fungi are important for mine remediation and rehabilitation, such as Glomus sp. in Veronica rechingeri growing in Pb/Zn mine, Glomus intraradices in Zea mays and Plantago lanceolata growing in Pb/Zn mine as well as Glomus etunicatum in Salsola kali and Panicum virgatum growing in an iron mine site.

Carbon Sequestration

Carbon (C) sequestration in soil refers to transferring CO_2 from the atmosphere into pedological and plant C pool in the form of soil inorganic carbon (SIC) that includes elemental C and carbonate minerals ($CaCO_3$ and $MgCO_3$), and organic carbon (SOC) which is stored in plants and organic matter (humus). Benefits of terrestrial C sequestration include: (a) Mitigation of climatic changes reducing net increase in atmospheric CO_2 concentration; (b) restoring soil quality; (c) improving water and nutrient storage capacity and elemental recycling; (d) reducing risk of erosion; (e) improving the environment; (f) increasing biodiversity. The SOC pool is the largest terrestrial source of carbon to 3 m depth. A positive SOC budget can be achieved by increasing the input of aboveground (leaves, stems, branches; litter and understory vegetation) and belowground (roots including root exudates) biomass and use of organic fertilizers (compost and manure), and decreasing the loss or output of carbon by erosion, leaching, oxidation, and mineralization. An important component of the global carbon cycle is soil CO_2 flux that represents soil respiration which includes respiration from roots and microbial respiration.

Revegetation of fly ash and reclaimed mine land increases SOC creating large carbon sink capacity which can be used as a potential pool for carbon sequestration. According to Pandey et al. the fly ash CO_2 flux from vegetated site was higher than from non-vegetated site. However, the CO_2 efflux rates were low in Saccharum spontaneum (84.29%) and Prosopis juliflora (92.09%) associations released minimum amount of CO_2 gas into the atmosphere leading to higher CO_2 sequestration. In the same time, the S. spontaneum had the highest aboveground and belowground biomass, and high content of organic carbon which can prove that this species is suitable for carbon sequestration. Furthermore, showed that in 11-years old reclaimed coal mine site CO_2 flux was lower (2.4 μmol CO_2 $m^{-2}s^{-1}$) than in reference to forest soil (3.8 μmol CO_2 $m^{-2}s^{-1}$) due to lower biomass and SOC pool. The CO_2 flux from forest soil in China and Europe was higher than reclaimed coal mine. However, over time, more carbon is accumulated in plant biomass where in turn more CO_2 is sequesterd presenting a win-win option in revegetation planning and phytomanagement of reclaimed sites.

Micoremediation

Endophytic non-mycorhizal fungi can promote the remediation of degraded soils due to their regulation of metal(loid)s uptake. Therefore, root-colonizing dark septate endophytic fungi (DSE) present conidial or sterile ascomycetes that may occur in contaminated sites protecting plants from meta and Hymenoscphus ericae in Pinus. In addition, DSE are involved in nutrient cycling i.e., they colonize older parts of the roots where nutrients are redistributed from senescent roots into the active roots. Furthermore, melanin in the cell wall of DSE hyphae may decrease metal(loid)s toxicity (Cu, Zn, Pb, Cd) increasing the activity of antioxidant enzymes (SOD, CAT) which in turn decreases oxidative stress.

Plant Adaptive Response to Envirionmental Stress

Integration of plant physiology and ecological restoration provides new perspectives on the relationship between plant physiological response to environmental stress and restoration ecology. International guideline of the Society for Ecological Restoration provides many potential opportunities for using physiological tools to evaluate the success of restoration plans and projects. Ecophysiological models can be important for understanding processes between environmental stress factors and adaptive response of plants which may help managers for decision-making and assessment of success of different restoration strategies. In addition, the fundamental understanding of plant physiology and biochemistry is important for selection of appropriate plant species for restoration of degraded sites.

Plant species that grow in the conditions of environmental stresses, such as drought, intense light, high temperature, salt stress, and toxicity/deficiency of metal(loids) show great variation in their mechanisms of tolerance. Environmental stressors affect photosynthesis, respiration, water regime and mineral nutrition of plants leading to the production of reactive oxygen species (ROS) causing "oxidative stress". The maintenance of cell redox homeostasis is achieved by activating the antioxidant plant system that presents a complex of control systems that enable them to overcome stress with minimal damage to the cell structure and function. The balance between ROS production and biosynthesis of antioxidants is very important in controlling redox homeostasis in plant cells. Therefore, the efficient removal of ROS can be achieved by activating the non-enzymatic and enzymatic components of antioxidant defense system. The non-enzymatic antioxidants include: carotenoids, α–tocopherol, phenolics, prolin, ascorbic acid (AsA), glutathione (GSH), phytochelatins (PCs), metallothionine (MTs). Enzymatic antioxidants include: superoxide dismutase (SOD), ascorbate peroxidase (APX), glutathione reductase (GR), glutathione peroxidase (GPX), catalase (CAT), glutathione S transferase (GST). However, ROS can act as signaling molecules in their adaptive response to stress. Thus, malondialdehyde (MDA) as a product of lipid peroxidation of membranes can act as the buffer of ROS with protective function i.e., it can be a signal for the gene expression of antioxidant enzymes and molecules during the oxidative stress. Generally, plants are well-adapted to the decreased amount of particular antioxidants inducing a signal for biosynthesis of other components of protection systems.

Plant Physiological Adaptation on Fly Ash

The photosynthetic efficiency of PSII (Fv/Fm) of plants growing on fly ash deposits were below the optimal range for plants which indicates a low number of functionally active PSII reaction centers, photoinhibition of PSII and lower vitality of plants in stress conditions. Low values of Fv/Fm were measured in herbaceous plants grown on fly ash deposits, such as Cirsium arvense, Epilobium collinum, Crepis bienis, Eupatorium cannabinum, Verbascum phlomoides, Calamagrostis epigejos, Oenothera biennis, Festuca rubra (0.429–0.691), as well as in shrubs, such as Tamarix tetandra, Amorpha fruticosa, Spiraea van-houttei (0.588–0.727), and trees Populus alba, Robinia pseudoaccacia (0.541–0.626). However, Festuca rubra growing on fly ash lagoons that are 11 years old showed increased photosynthetic efficiency (Fv/Fm) indicating that this sown species over time activates adaptive mechanisms that increase the tolerance of the photosynthetic apparatus to stress. Furthermore, plants growing on fly ash deposits showed a high ability to maintain a relatively favorable water balance. Thus, relative water content (RWC) in Spiraea van-houttei, Cirsium arvense, Verbascum phlomoide, Crepis setosa, Epilobium collinum, and Eupatorium cannabinumwere in the range 84.03–95.52%. However, low values of RWC (75.89%) and high values of water saturation deficit (WSD) (16.40%) were found in Populus alba. The sap osmotic potential was the most favorable in Crepis setosa, Cirsium arvense and Verbascum phlomoides (from −0.981 MPa to −1.135 MPa) whereas in Populus alba unfavorable sap osmotic value was detected (−1.377 MPa).

Decrease of chlorophylls and carotenoids content in plants growing on fly ash deposits were noted in Ricinus communis, Typha latifolia, Thelypteris dentate, Festuca rubra, Amorpha fruticosa, Tamarix tetandra, Populus alba and Robinia pseudoaccaci in relation to control. Furthermore, in Cicer arietinum which grew in fly ash treated soils, the content of chlorophylls decreased at 50 and 100% fly ash whereas carotenoids amount increased. Similarly, chlorophyls and carotenoids content decreased in leaves of Lolium perenne growing on the soil where fly ash was added. Small content of chlorophylls and carotenoids in plant leaves indicates that the photosynthetic pigments are sensitive to stress factors that prevail on fly ash, such as high temperature, drought, toxic concentrations of As, B, Cd, Cr, Pb, and the small amount of Cu, Mn, and Zn in leaves. However, displacement of Mg in the structure of chlorophyll molecule with toxic metal(loid) ions due to similar chemical properties reduces its activity and leads to the non-functional molecule. In addition, small content of carotenoids in plants may suggest decreased photoprotection of the PSII which in turn leads to the reduction of photosynthetic efficiency. Malondialdehyde (MDA) as an indicator of free radical production and lipid peroxidation of membranes indicates "oxidative stress" in plants. It was noted that MDA content in the roots and leaves of Cicer arietinum growing at 50 and 100% fly ash was high as well as in the roots and fronds of Thelypteris dentate at 75 and 100% fly ash. According to Gajić et al. in the leaves of Festuca rubra grown on fly ash deposits, the content of MDA was high probably as a result of toxic concentrations of As and B and the low content of Cu, Zn, and Mn.

Plant species that grow in the stress conditions prevailing on fly ash promote biosynthesis of antioxidants which can reduce the ROS leading to an increase in photosynthetic efficiency. Thus, in the leaves of sown species Festuca rubra and spontaneously colonized species Calamagrostis epigejosand Oenothera biennis growing on fly ash deposits, increased content of phenolics, ascorbic acids and radical scavenging activity were noted. Furthermore, increased activity of superoxide dismutase (SOD), catalases (CAT), peroxidases (POD), glutathione reductase (GR), and glutathione S-transferase (GST) in Cicer arietinumgrowing at 50% and 100% fly ash were noted as well as increased content of cysteine and non-protein thiols (NP-SH) in Thelypteris dentate and Oryza sativa. In experimental conditions with different levels of fly ash application, content of ascorbic acids, phenolics, prolin, cysteine, and NP-SH in the leaves of Beta vulgaris and Sesbania cannabina increased. Increased biosynthesis of antioxidants in plant species that can grow on fly ash indicates that they possess a high ecophysiological adaptive potential to endure harsh conditions prevailing on fly ash deposits.

Plant Physiological Adaptation on Mine Waste

Plant physiological and biochemical response to multiple stresses on mine waste rock and tailings may indicate their capacity to thrive under extreme environmental conditions. The knowledge related to plant metabolite response can be useful for mine land reclamation programs. Thus, changes in antioxidative/oxidative parameters in Zygophylum fabago growing on mine tailings revealed significant decrease of chlorophylls content, reduction of MDA content, H_2O_2 and O_2^- levels, proline content and the ionically-bound cell wall peroxidases (iPRX) whereas the total antioxidant capacity, particularly content of phenolic compounds (flavonoids, proanthocyanidins, hydroxycinnamic acids), ascorbic acid, GSH, NPT (total non-protein thiol), soluble sugars and activity of PAL (phenylalanine ammonia lyase) as well as activity of soluble peroxidases (sPRX) significantly increased. Therefore, reduction of oxidative stress is associated with the activation of plant antioxidant system to cope with unforable multi-element pollution and harsh climatic conditions on mine tailings. Similarly, in Dittrichia viscose growing on mine tailings, no changes were found in ROS, starch, soluble sugar concentrations, ascorbic acid and GSH content whereas the content of cell wall phenolics, PAL and sPRX activity increased. Slight reduction in the total antioxidant activity was related to the reduction of soluble phenolics.

However, coordination of phenolic compounds in phenypropanoid metabolisms could have a prominent role in the capability of plants to cope with oxidative stress and toxic pollutants in mine areas. Epilobium dodonaei growing on mine waste rock showed the reduction in chlorophylls and carotenoids amount and great content of MDA. Despite high oxidative stress, this species is capable to promote biosynthesis of soluble and cell-wall phenolics and other free radical scavenging compounds as a response to metal(loid)s stress on mine spoil. Lu et al. showed that Sonchus transcaspicus had a higher adaptive potential to metal-induced oxidative stress compared to Arrhenatherum elatium growing on mine tailings. Photosynthetic efficiency (Fv/Fm) and chlorophyll content decreased in both plants, but it is more pronounced in A. elatius. MDA content in the roots and leaves of A. elatius was high, whereas MDA content in the shoots of S. transcaspicus was not affected. Furthermore, in S. transcaspicus activity of SOD, POD, CAT, and APX in the shoots and roots was high. Activity of all these enzymes in the roots of A. elatius was also high, but in the shoots only POD activity was high. Nadgorska-Socha et al. found that in Cardaminopsis arenosa and Plantago lanceolata growing in mining areas, activity of SOD, POD, and GSH increased whereas non-protein—SH group content and proline decreased as a response to metal exposure.

Plant Biochemistry

Plant biochemistry involves the study of the biochemistry of autotrophic organisms such as photosynthesis and other biochemical processes specific to plants. Plant Biochemistry presents each topic from the cellular level to the ecological and environmental levels, placing it in the context of the whole plant. Biochemical pathways are represented as route maps, showing how one reaction follows another. These maps emphasize the dynamism and flexibility of the plant facing the environmental challenges. The unique and wide–ranging approach of this book emphasizes the importance of teaching and learning pathways within the framework of what the pathway does and why it is needed. Plant Biochemistry illustrates the impact of plants on human activity and success, in terms of their importance as a food supply and as raw materials for industrial and pharmaceutical products, and considers that humans can benefit from exploiting plant biochemical pathways. So, Plant Biochemistry is the science which studies chemical and physico–chemical processes that take place in the living organisms having the role to establish the material substrate of the life's phenomena. Usually the development of the organisms is possible due to the biochemical processes that take place in them in different environmental conditions. The influence exerted by the environment is reflected in the morphological and chemical structure of the organisms.

The fundamental problems of the animal biochemistry cannot be broken from the vegetal biochemistry and from the unicellular organisms. The universal principle of nature's integrity is the base of the chemical composition of the organisms and of the processes that are continuously developing and whose meaning is life. The organic matter is forming, transforming and degrading in any living body (plant, animal or microorganism); it doesn't differ from an organism to other by the component substances that are always the same or related also differs through secondary and unessential features determined by the proportion and the conditions of the formation and transformation of the organic matter. This is the reason why in the organisms' life there are some formation and degradation developing processes of the live matter and they vary between certain limits determining the variability and heredity of the organisms. Animal, plant and environment are forming an indissoluble unit and the biochemistry shows the diversity of the live matter.

Plant biochemistry is related to the animal and human activity where there are similar, however in all plants each cell has achieved a much more complicated metabolic process comparing to the animal organisms which are specialized on different functions. Animal, plant and environment form an inseparable unit, and only this way, in a comparing and general way, biochemistry shows a material world and the different shapes that it can take. So, when studying biochemistry there have to be treated at the same time both the fundamental chemistry of animals and that one of chemical plants and their need. Biochemistry has a particular role in the acquisition of agricultural raw materials for finished products with good yields, quality, and economic conditions in high yield. Biochemistry is involved in obtaining such materials in the conduct of technological processes, to conserve packaging and finished products, waste recovery in manufacturing, etc. During their industrial transformation raw materials and agricultural products suffer biochemical processes necessary or harmful, caused by enzymes, processes that must be traced, suppressed or directed.

Any living organism is an energetic system. The amount of energy received from the environment is equal to the consumed energy, plus the accumulations. If the accumulations are positive the organism will gain weight, and if the accumulations are negative it will lose weight. In a healthy organism it is desired that the accumulation should be maintained constant in time. The totality of the energetic exchanges that take place permanently between the living body and the environment, meaning that the organism is spending (losing) energy replaced by the energy contained in food is called energetic expenses.

The energy from the nutrients is the potential chemical energy and it is contained in the C–H, C–O, and the C–N links from proteins, lipids and carbohydrates, these links being able to split in the conditions of the high temperatures, the pH and the chemical environment created by cellular activities. After the usual oxidative splitting of the nutrients, the organism stocks the potential chemical energy and transforms the energy in other energy forms necessary to it. If the energetic needs of the organism are reduced and the energetic effort is increased, the glucose, the fatty acids and the amino acids are being entirely catabolised and transformed in stocking forms such as glycogen, triglucides and proteins.

The basis of all the biological processes is the exchange of matter, information and energy, this exchange process being active between the living body and the environment. The nutrients, the environmental factor indispensable for life, are crude or processed substances designated for the nutrition and satisfaction of the energy necessary for the maintaining the vital functions, for providing the growing and development of the living body and for maintaining the aptitudes for a good physical and mental activity.

The nutrients are complex products which are capable to suffer the necessary transformations in the living organism in order to maintain life and forming the tissues. Along with the oxygen, the living body also incorporates water and minerals, proteins, lipids, carbohydrates and vitamins from environment. Except oxygen, which is essential for the respiration process and the water, which is the vital environment, the other compounds are imported. The consumption of substances coming from external sources is a fundamental biological necessity of any organism. Other substances are named nutritive principles or alimentary principles and for their definition are used the terms nutrients or trophynes. All metabolic processes take place in cells where the mitochondria is the main centre of the energogenesis, and the ribosomes are the place for the proteins' synthesis.

The functions of the nutritive principles found in food are:

- Catalytic favouring the normal development of the biological processes.
- Energetic providing the necessary energy for the development of the vital processes.
- Plastic providing the synthesis of the substances owned by the organism; etc.

Biochemistry studies the process of the incorporation of the substances which participate in the metabolic processes, the transformation of the nutrients in soluble substances capable to be absorbed by the blood and produced in the digestive apparatus. Food's nutritive factors have some complex combinations that cannot be used and necessitate their transformation in simple elements which can be easily assimilated by the organism.

The Nutritive Principles and Fundamental Substances

Modern biochemistry agrees with the existence of nine categories of nutritive principles: Water, alimentary fibreslipids, carbohydrates, macrominerals, oligominerals, proteins, bioactive substances and vitamins.

The chemical substances which are part of these categories have precise functions useful to the organism that ingrates them. Thus some of them provide energy and heat, others provide construction material for the organism helping with the growth or the repair of the degraded tissues, and others participate directly or indirectly in the regulation of the biochemical and physiological processes from the organism. Yet none of these nutritive principles activate independently. For a good metabolic activity and health of the organism the alimentation should contain all those nine nutritive principles in adequate amounts. Due to their structure and properties the organic substances are the basic material of the living bodies and activate in all living cells. The organic compounds mainly consist of: carbon, hydrogen, oxygen, nitrogen, phosphorus and sulphur. Proteins, lipids and carbohydrates are part of this category and serve as plastic and energetic material with a distinct structural and functional role in the living bodies. The living matter is formed of two main groups of substances: inorganic and organic. The inorganic substances (water, minerals) are up taken from environment and are not synthesized by plants. The organic substances are the main compounds of the living bodies. According to their role in organism they can be classified as:

- Plastic substances: (Carbohydrates, proteins, and lipids) with a structural and energetic role.
- Substances with informational role: Nucleic acids.
- Substances with catalytic role: Enzymes.
- Substances with a role in balance: Enzymes, vitamins, phytohormons, pigments, antibodies etc.
- Substances with an energetic role: Macroergic compounds.

Intermediary and final substances of the metabolism: Glycosides, lignin, eteric oils, alkaloids, antibiotic substances, alcohols, amides, amines, esters etc.

The living matter is formed especially of the compounds of 12 chemical elements called plastic macroelements or constitutive elements. These are: C, H, O, N, P, S, Cl, K, Na, Ca, Mg, and Si. These plastic elements, together with other elements a.i. oligo elements or microelements: Mo, Cu, Zn, Fe, B, Mn, Co, As, F, Br, Al etc. make less than 1% of the dry matter. The main source of supply with inorganic compounds for plants is the soil. Mineral substances, component part of the food, are important in forming tissues and participate in the biological and physiological processes of the body. They are classified into macroelements (K, Ca, P, Na, Mg, Fe, Cl, S), which exist in the body in large amounts and microelements (Zn, Cu, Cr, Mn, Co, Mo, I, F, Ni etc.) found in small quantities. The microelements have a very important functional role in the reduction reactions, in the balance of the osmotic pressure and of the acid–base balance, and in maintaining cellular integrity even if they are found in small amount. Having also an essential role in the activation of some enzymes they become catalytic elements. Studies show the presence in the vegetal organisms of more than 60 chemical elements.

From these only seven are fundamental or macroelements and together with the C, H, O and N they form about 99% of the body weight. These macroelements are: Na, K, Ca, P, Mg, Cl, and S. The other mineral elements found in very small amount are called microelements. The most important microelements are: Fe, Cu, Co, Zn, I, Mn, Cd, Se, Nil, Cr, Pb, Hg, and As. Macro and microelements are indispensable for the physiological equilibrium of the vegetal, animal and human organisms and the lack of one or more minerals results in pathological manifestations more or less severe.

Carbon is part of those organic compounds existing in the organisms and represents 48–50% of the dry substance. Hydrogen represents 6–7% of the dry matter and just as carbon, is found in all organic substances and in the water from body. It also participates in the oxidation–reduction reactions. Oxygen is found in most of the organic compounds from the living bodies in a proportion of 41–43% and along with the hydrogen, takes part in the oxidatio–reduction processes. Nitrogen is one of the basic elements of protides, and represents 16% of plant protides. In living tissues the proportion of nitrogen reaches 1.4–1.6%. Phosphorus is part of some important biologic compounds such as: enzymes, complex lipids, ATP etc. and interferes in many biochemical processes (photosynthesis etc.). In living bodies is found in a proportion near to 0.1–1.1%. Sulphur is part of some proteins, vitamins and carbohydrates etc. and can be found as 0.02–1.8% of the dry matter. It has important role in the growth and development of the living bodies.

The basic components in the alimentation of living organisms are the essential amino acids, polyunsaturated fatty acids, all the vitamins and minerals. Some components with high physiological activity, such as phospholipids, glycoproteidele, phosphoproteinele and many other compounds, are considered essential substances. The basic principles of a balanced nutrition are the proper correlations based on substances and biologically active nutrients, like proteins, lipids, carbohydrates, vitamins, mineral substances, depending on age, gender, activity at work and lifestyles. Nutritional factors are subjected to a series of changes before their addition to the body tissues i.e. the burning and the disposal of unused remainders, formed as a result of metabolic processes taking place in the body.

The metabolism represents the chemical and energetic changes of the products resulted from digestion and which are necessary in the anabolic and catabolic processes in the organism. The anabolism is the synthesis and repair process of simple substances' tissues. The catabolism (decomposition, degradation) is a decomposition process of the composed substances used as an energy source.

Photosynthesis

Photosynthesis is the ultimate source of all of humankind's food and oxygen, whereas fossilized photosynthetic fuels provide ~87% of the world's energy. It is the biochemical process that sustains the biosphere as the basis for the food chain. The oxygen produced as a by-product of photosynthesis allowed the formation of the ozone layer, the evolution of aerobic respiration and thus complex multicellular life.

Oxygenic photosynthesis involves the conversion of water and CO_2 into complex organic molecules such as carbohydrates and oxygen. Photosynthesis may be split into the 'light' and 'dark' reactions. In the light reactions, water is split using light into oxygen, protons and electrons, and in the dark reactions, the protons and electrons are used to reduce CO_2 to carbohydrate (given here by the general formula CH_2O). The two processes can be summarized thus:

Light reactions:

$$2H_2O + \text{light} \rightarrow O_2 + 4H^+ + 4e^- \left(\Delta G^\circ = +317\ \text{kJ} \cdot \text{mol}^{-1}\right)$$

Dark reactions:

$$CO_2 + 4H^+ + 4e^- \rightarrow CH_2O + H_2O\left(\Delta G^\circ = +162\ \text{kJ} \cdot \text{mol}^{-1}\right)$$

Overall:

$$H_2O + \text{light} + CO_2 \rightarrow CH_2O + O_2 \left(\Delta G^\circ = +479\ \text{kJ} \cdot \text{mol}^{-1}\right)$$

The positive sign of the standard free energy change of the reaction (ΔG^o) given above means that the reaction requires energy (an endergonic reaction). The energy required is provided by absorbed solar energy, which is converted into the chemical bond energy of the products.

Photosynthesis converts ~200 billion tonnes of CO_2 into complex organic compounds annually and produces ~140 billion tonnes of oxygen into the atmosphere. By facilitating conversion of solar energy into chemical energy, photosynthesis acts as the primary energy input into the global food chain. Nearly all living organisms use the complex organic compounds derived from photosynthesis as a source of energy. The breakdown of these organic compounds occurs via the process of aerobic respiration, which of course also requires the oxygen produced by photosynthesis.

$$CH_2O + O_2 \rightarrow CO_2 + H_2O(\Delta G^\circ = -479\ \text{kJ} \cdot \text{mol}^{-1})$$

Unlike photosynthesis, aerobic respiration is an exergonic process (negative ΔG^o) with the energy released being used by the organism to power biosynthetic processes that allow growth and renewal, mechanical work (such as muscle contraction or flagella rotation) and facilitating changes in chemical concentrations within the cell (e.g. accumulation of nutrients and expulsion of waste). The use of exergonic reactions to power endergonic ones associated with biosynthesis and housekeeping in biological organisms such that the overall free energy change is negative is known as 'coupling'.

Photosynthesis and respiration are thus seemingly the reverse of one another, with the important caveat that both oxygen formation during photosynthesis and its utilization during respiration result in its liberation or incorporation respectively into water rather than CO_2. In addition, glucose is one of several possible products of photosynthesis with amino acids and lipids also being synthesized rapidly from the primary photosynthetic products.

The consideration of photosynthesis and respiration as opposing processes helps us to appreciate their role in shaping our environment. The fixation of CO_2 by photosynthesis and its release during breakdown of organic molecules during respiration, decay and combustion of organic matter and fossil fuels can be visualized as the global carbon cycle.

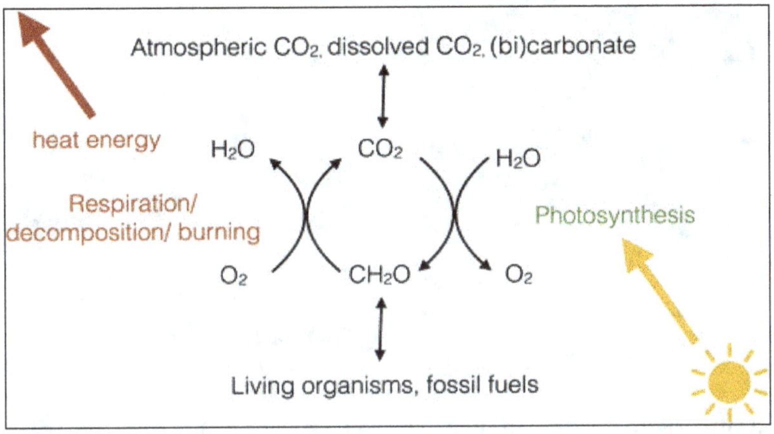

The relationship between respiration, photosynthesis and global CO_2 and O_2 levels.

At present, this cycle may be considered to be in a state of imbalance due to the burning of fossil fuels (fossilized photosynthesis), which is increasing the proportion of CO_2 entering the Earth's atmosphere, leading to the so-called 'greenhouse effect' and human-made climate change.

Oxygenic photosynthesis is thought to have evolved only once during Earth's history in the cyanobacteria. All other organisms, such as plants, algae and diatoms, which perform oxygenic photosynthesis actually do so via cyanobacterial endosymbionts or 'chloroplasts'. An endosymbiotic event between an ancestral eukaryotic cell and a cyanobacterium that gave rise to plants is estimated to have occurred ~1.5 billion years ago. Free-living cyanobacteria still exist today and are responsible for ~50% of the world's photosynthesis. Cyanobacteria themselves are thought to have evolved from simpler photosynthetic bacteria that use either organic or inorganic compounds such a hydrogen sulfide as a source of electrons rather than water and thus do not produce oxygen.

The Site of Photosynthesis in Plants

In land plants, the principal organs of photosynthesis are the leaves. Leaves have evolved to expose the largest possible area of green tissue to light and entry of CO_2 to the leaf is controlled by small holes in the lower epidermis called stomata. The size of the stomatal openings is variable and regulated by a pair of guard cells, which respond to the turgor pressure (water content) of the leaf, thus when the leaf is hydrated, the stomata can open to allow CO_2 in. In contrast, when water is scarce, the guard cells lose turgor pressure and close, preventing the escape of water from the leaf via transpiration.

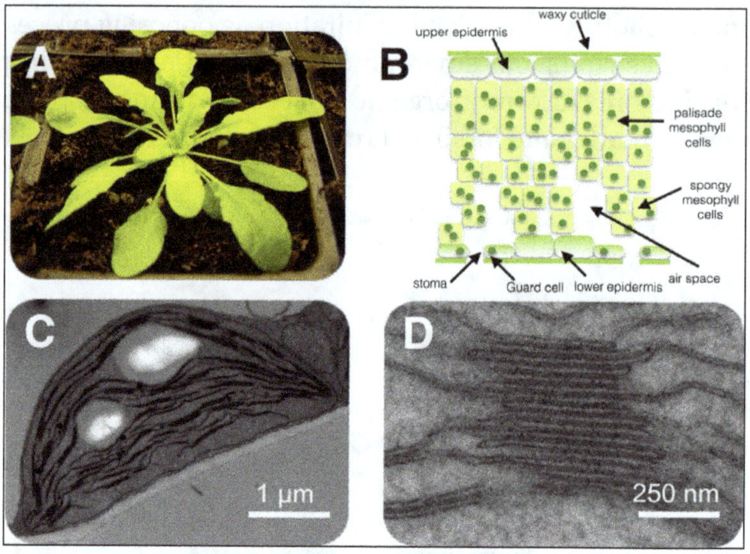

Location of the photosynthetic machinery: (A) The model plant *Arabidopsis thaliana*. (B) Basic structure of a leaf shown in cross-section. Chloroplasts are shown as green dots within the cells. (C) An electron micrograph of an *Arabidopsis* chloroplast within the leaf. (D) Close-up region of the chloroplast showing the stacked structure of the thylakoid membrane.

Within the green tissue of the leaf (mainly the mesophyll) each cell (~100 µm in length) contains ~100 chloroplasts (2–3 µm in length), the tiny organelles where photosynthesis takes place. The chloroplast has a complex structure with two outer membranes (the envelope), which are colourless and do not participate in photosynthesis, enclosing an aqueous space (the stroma) wherein sits a third membrane known as the thylakoid, which in turn encloses a single continuous aqueous space called the lumen.

The light reactions of photosynthesis involve light-driven electron and proton transfers, which occur in the thylakoid membrane, whereas the dark reactions involve the fixation of CO_2 into carbohydrate, via the Calvin–Benson cycle, which occurs in the stroma. The light reactions involve electron transfer from water to $NADP^+$ to form NADPH and these reactions are coupled to proton transfers that lead to the phosphorylation of adenosine diphosphate (ADP) into ATP. The Calvin–Benson cycle uses ATP and NADPH to convert CO_2 into carbohydrates, regenerating ADP and $NADP^+$. The light and dark reactions are therefore mutually dependent on one another.

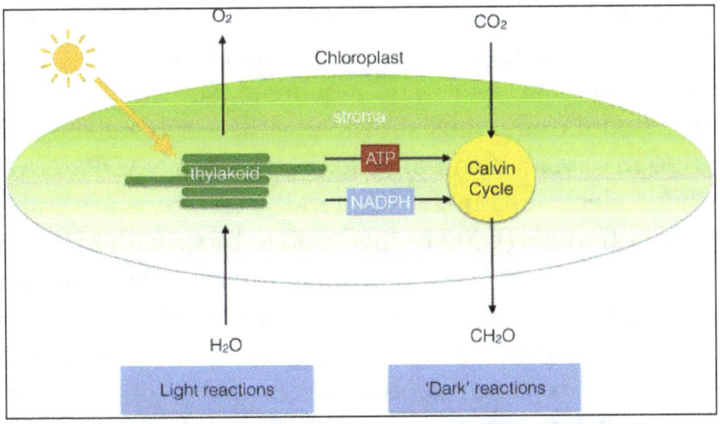

Division of labour within the chloroplast. The light reactions of photosynthesis take place in the thylakoid membrane, whereas the dark reactions are located in the chloroplast stroma.

Photosynthetic Electron and Proton Transfer Chain

The light-driven electron transfer reactions of photosynthesis begin with the splitting of water by Photosystem II (PSII). PSII is a chlorophyll–protein complex embedded in the thylakoid membrane that uses light to oxidize water to oxygen and reduce the electron acceptor plastoquinone to plastoquinol. Plastoquinol in turn carries the electrons derived from water to another thylakoid-embedded protein complex called cytochrome b_6f (cytb_6f). cytb_6f oxidizes plastoquinol to plastoquinone and reduces a small water-soluble electron carrier protein plastocyanin, which resides in the lumen. A second light-driven reaction is then carried out by another chlorophyll protein complex called Photosystem I (PSI). PSI oxidizes plastocyanin and reduces another soluble electron carrier protein ferredoxin that resides in the stroma. Ferredoxin can then be used by the ferredoxin–NADP$^+$ reductase (FNR) enzyme to reduce NADP$^+$ to NADPH. This scheme is known as the linear electron transfer pathway or Z-scheme.

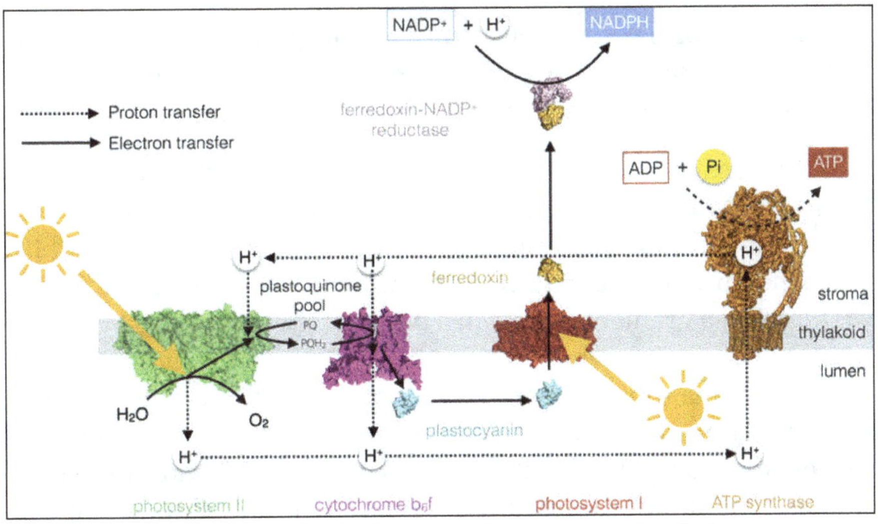

The photosynthetic electron and proton transfer chain.

The linear electron transfer pathway from water to $NADP^+$ to form NADPH results in the formation of a proton gradient across the thylakoid membrane that is used by the ATP synthase enzyme to make ATP. The Z-scheme, so-called since it resembles the letter 'Z' when turned on its side, thus shows how the electrons move from the water–oxygen couple (+820 mV) via a chain of redox carriers to $NADP^+$/NADPH (−320 mV) during photosynthetic electron transfer. Generally, electrons are transferred from redox couples with low potentials (good reductants) to those with higher potentials (good oxidants) (e.g. during respiratory electron transfer in mitochondria) since this process is exergonic. However, photosynthetic electron transfer also involves two endergonic steps, which occur at PSII and at PSI and require an energy input in the form of light. The light energy is used to excite an electron within a chlorophyll molecule residing in PSII or PSI to a higher energy level; this excited chlorophyll is then able to reduce the subsequent acceptors in the chain. The oxidized chlorophyll is then reduced by water in the case of PSII and plastocyanin in the case of PSI.

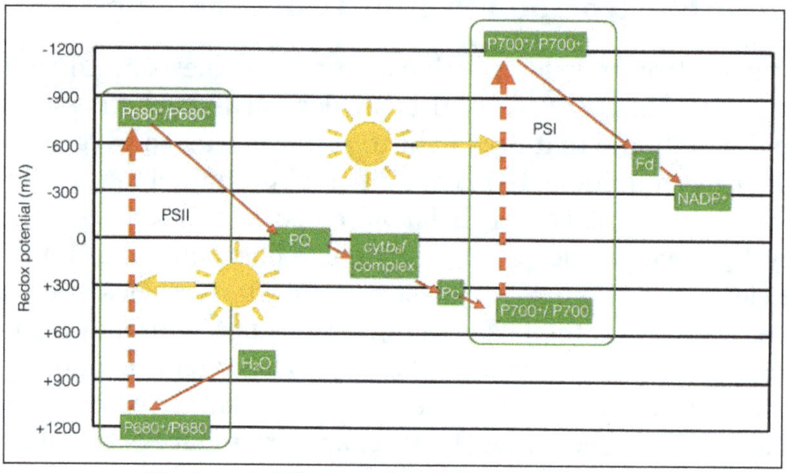

Z-scheme of photosynthetic electron transfer.

The main components of the linear electron transfer pathway are shown on a scale of redox potential to illustrate how two separate inputs of light energy at PSI and PSII result in the endergonic transfer of electrons from water to $NADP^+$.

The water-splitting reaction at PSII and plastoquinol oxidation at cytb_6f result in the release of protons into the lumen, resulting in a build-up of protons in this compartment relative to the stroma. The difference in the proton concentration between the two sides of the membrane is called a proton gradient. The proton gradient is a store of free energy (similar to a gradient of ions in a battery) that is utilized by a molecular mechanical motor ATP synthase, which resides in the thylakoid membrane. The ATP synthase allows the protons to move down their concentration gradient from the lumen (high H^+ concentration) to the stroma (low H^+ concentration). This exergonic reaction is used to power the endergonic synthesis of ATP from ADP and inorganic phosphate (P_i). This process of photophosphorylation is thus essentially similar to oxidative phosphorylation, which occurs in the inner mitochondrial membrane during respiration.

An alternative electron transfer pathway exists in plants and algae, known as cyclic electron flow. Cyclic electron flow involves the recycling of electrons from ferredoxin to plastoquinone, with the result that there is no net production of NADPH; however, since protons are still transferred into the lumen by oxidation of plastoquinol by cytb_6f, ATP can still be formed. Thus photosynthetic organisms can control the ratio of NADPH/ATP to meet metabolic need by controlling the relative amounts of cyclic and linear electron transfer.

How the Photosystems Work

Light Absorption by Pigments

Photosynthesis begins with the absorption of light by pigments molecules located in the thylakoid membrane. The most well-known of these is chlorophyll, but there are also carotenoids and, in cyanobacteria and some algae, bilins. These pigments all have in common within their chemical structures an alternating series of carbon single and double bonds, which form a conjugated system π–electron system.

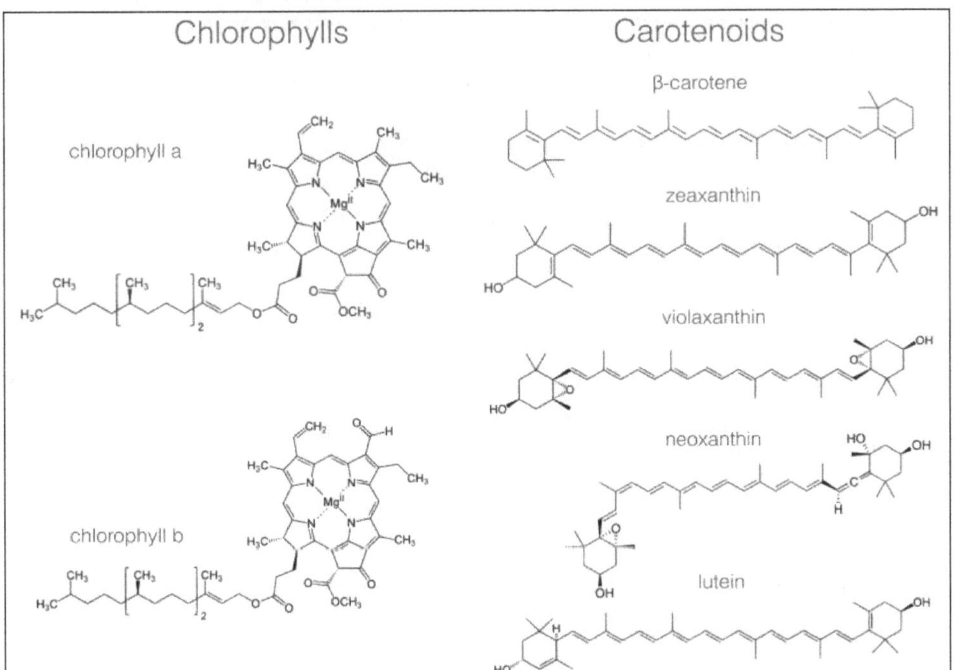

Major photosynthetic pigments in plants.

The chemical structures of the chlorophyll and carotenoid pigments present in the thylakoid membrane. Note the presence in each of a conjugated system of carbon–carbon double bonds that is responsible for light absorption.

The variety of pigments present within each type of photosynthetic organism reflects the light environment in which it lives; plants on land contain chlorophylls *a* and *b* and carotenoids such as β-carotene, lutein, zeaxanthin, violaxanthin, antheraxanthin and neoxanthin. The chlorophylls absorb blue and red light and so appear green in colour, whereas carotenoids absorb light only in the blue and so appear yellow/red, colours more obvious in the autumn as chlorophyll is the first pigment to be broken down in decaying leaves.

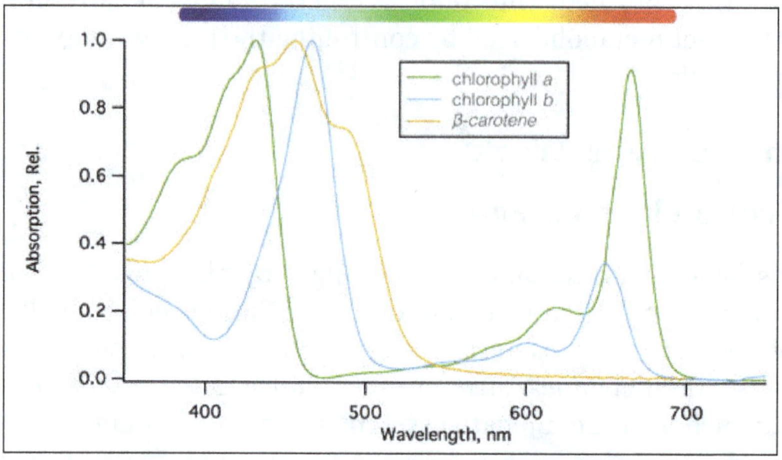

Basic absorption spectra of the major chlorophyll and carotenoid pigments found in plants. Chlorophylls absorb light energy in the red and blue part of the visible spectrum, whereas carotenoids only absorb light in the blue/green.

Light, or electromagnetic radiation, has the properties of both a wave and a stream of particles (light quanta). Each quantum of light contains a discrete amount of energy that can be calculated by multiplying Planck's constant, h (6.626×10^{-34} J·s) by ν, the frequency of the radiation in cycles per second (s^{-1}):

$$E=h\nu$$

The frequency (ν) of the light and so its energy varies with its colour, thus blue photons (~450 nm) are more energetic than red photons (~650 nm). The frequency (ν) and wavelength (λ) of light are related by:

$$\lambda=c/\nu$$

where c is the velocity of light (3.0×10^{8} m·s^{-1}), and the energy of a particular wavelength (λ) of light is given by:

$$E=hc/\lambda$$

Thus 1 mol of 680 nm photons of red light has an energy of 176 kJ·mol^{-1}.

The electrons within the delocalized π system of the pigment have the ability to jump up from the lowest occupied molecular orbital (ground state) to higher unoccupied molecular electron orbitals (excited states) via the absorption of specific wavelengths of light in the visible range (400–725 nm). Chlorophyll has two excited states known as S_1 and S_2 and, upon interaction of the molecule with a photon of light, one of its π electrons is promoted from the ground state (S_0) to an excited state, a process taking just 10^{-15} s. The energy gap between the S_0 and S_1 states is spanned by the energy provided by a red photon (~600–700 nm), whereas the energy gap between the S_0 and S_2 states is larger and therefore requires a more energetic (shorter wavelength, higher frequency) blue photon (~400–500 nm) to span the energy gap.

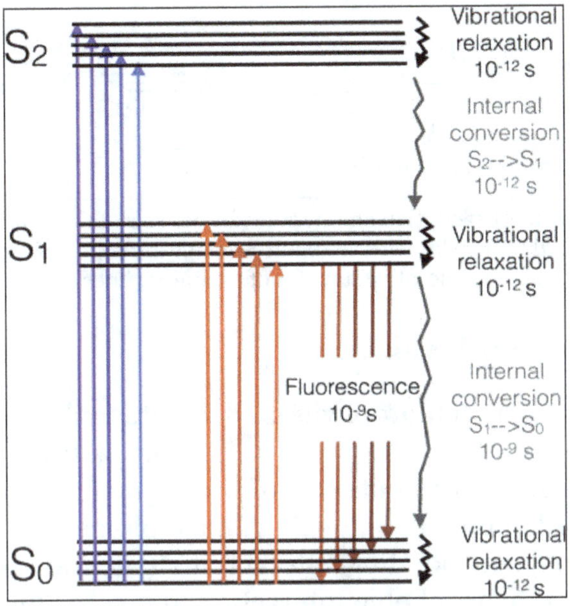

Diagram of chlorophyll showing the possible fates of the S_1 and S_2 excited states and timescales of the transitions involved. Photons with slightly different energies (colours) excite each of the vibrational substates of each excited state (as shown by variation in the size and colour of the arrows).

Upon excitation, the electron in the S_2 state quickly undergoes losses of energy as heat through molecular vibration and undergoes conversion into the energy of the S_1 state by a process called internal conversion. Internal conversion occurs on a timescale of 10^{-12} s. The energy of a blue photon is thus rapidly degraded to that of a red photon. Excitation of the molecule with a red photon would lead to promotion of an electron to the S_1 state directly. Once the electron resides in the S_1 state, it is lower in energy and thus stable on a somewhat longer timescale (10^{-9} s). The energy of the excited electron in the S_1 state can have one of several fates: it could return to the ground state (S_0) by emission of the energy as a photon of light (fluorescence), or it could be lost as heat due to internal conversion between S_1 and S_0. Alternatively, if another chlorophyll is nearby, a process known as excitation energy transfer (EET) can result in the non-radiative exchange of energy between the two molecules. For this to occur, the two chlorophylls must be close by (<7 nm), have a specific orientation with respect to one another, and excited state energies that overlap (are resonant) with one another. If these conditions are met, the energy is exchanged, resulting in a mirror $S_0 \rightarrow S_1$ transition in the acceptor molecule and a $S_1 \rightarrow S_0$ transition in the other.

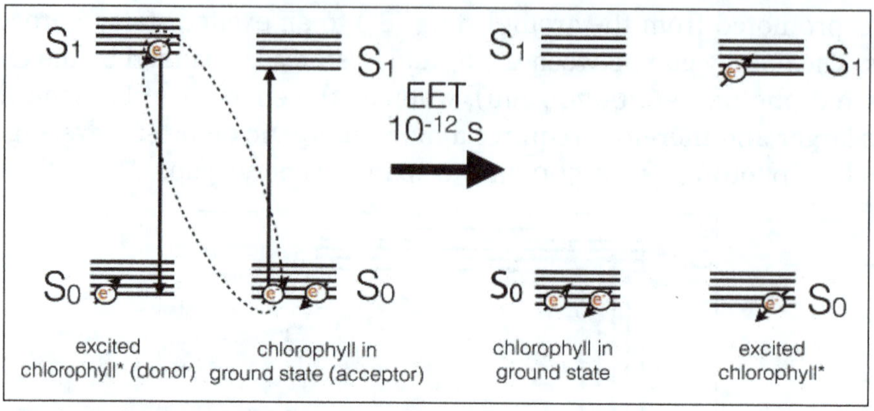

Basic mechanism of excitation energy transfer between chlorophyll molecules. Two chlorophyll molecules with resonant S_1 states undergo a mirror transition resulting in the non-radiative transfer of excitation energy between them.

Light-harvesting Complexes

In photosynthetic systems, chlorophylls and carotenoids are found attached to membrane-embedded proteins known as light-harvesting complexes (LHCs). Through careful binding and orientation of the pigment molecules, absorbed energy can be transferred among them by EET. Each pigment is bound to the protein by a series of non-covalent bonding interactions (such as, hydrogen bonds, van der Waals interactions, hydrophobic interaction and co-ordination bonds between lone pair electrons of residues such as histidine in the protein and the Mg^{2+} ion in chlorophyll); the protein structure is such that each bound pigment experiences a slightly different environment in terms of the surrounding amino acid side chains, lipids, etc., meaning that the S_1 and S_2 energy levels are shifted in energy with respect to that of other neighbouring pigment molecules. The effect is to create a range of pigment energies that act to 'funnel' the energy on to the lowest-energy pigments in the LHC by EET.

Reaction Centres

A photosystem consists of numerous LHCs that form an antenna of hundreds of pigment molecules. The antenna pigments act to collect and concentrate excitation energy and transfer it towards a 'special pair' of chlorophyll molecules that reside in the reaction centre (RC). Unlike the antenna pigments, the special pair of chlorophylls are 'redox-active' in the sense that they can return to the ground state (S_0) by the transfer of the electron residing in the S_1 excited state (Chl*) to another species. This process is known as charge separation and result in formation of an oxidized special pair (Chl^+) and a reduced acceptor (A^-). The acceptor in PSII is plastoquinone and in PSI it is ferredoxin. If the RC is to go on functioning, the electron deficiency on the special pair must be made good, in PSII the electron donor is water and in PSI it is plastocyanin.

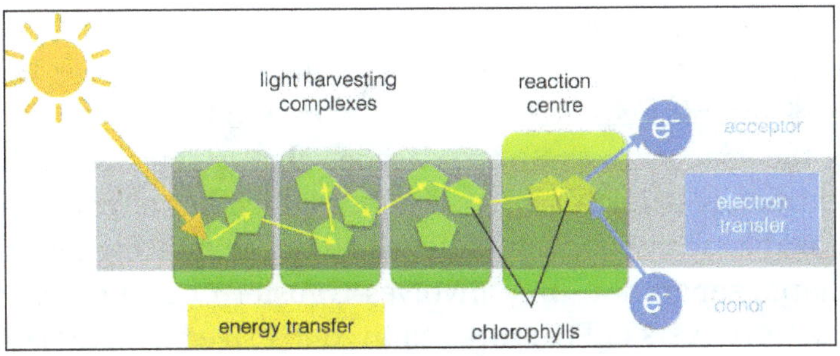

Basic structure of a photosystem.

Light energy is captured by the antenna pigments and transferred to the special pair of RC chlorophylls which undergo a redox reaction leading to reduction of an acceptor molecule. The oxidized special pair is regenerated by an electron donor.

It is worth asking why photosynthetic organisms bother to have a large antenna of pigments serving an RC rather than more numerous RCs. The answer lies in the fact that the special pair of chlorophylls alone have a rather small spatial and spectral cross-section, meaning that there is a limit to the amount of light they can efficiently absorb. The amount of light they can practically absorb is around two orders of magnitude smaller than their maximum possible turnover rate. Thus LHCs act to increase the spatial (hundreds of pigments) and spectral (several types of pigments with different light absorption characteristics) cross-section of the RC special pair ensuring that its turnover rate runs much closer to capacity.

Photosystem II

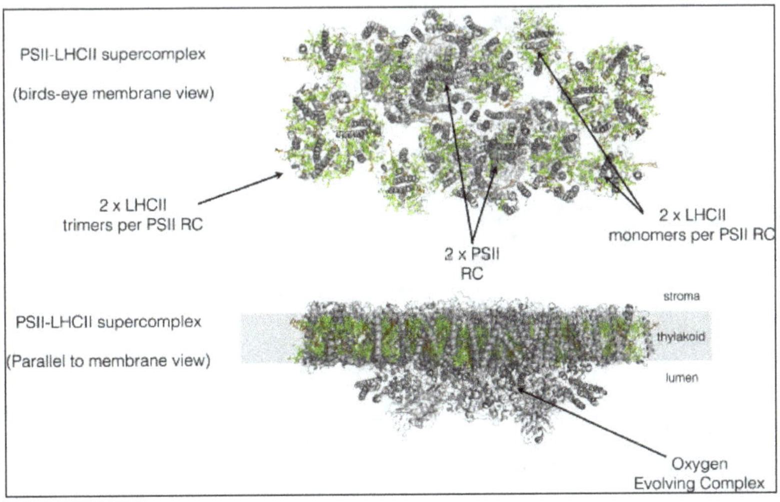

Basic structure of the PSII–LHCII supercomplex from spinach. The organization of PSII and its light-harvesting antenna. Protein is shown in grey, with chlorophylls in green and carotenoids in orange. Drawn from PDB code 3JCU.

PSII is a light-driven water–plastoquinone oxidoreductase and is the only enzyme in Nature that is capable of performing the difficult chemistry of splitting water into protons, electrons and oxygen. In principle, water is an extremely poor electron donor since the redox potential of the water–oxygen couple is +820 mV. PSII uses light energy to excite a special pair of chlorophylls, known as P680 due to their 680 nm absorption peak in the red part of the spectrum. P680* undergoes charge separation that results in the formation of an extremely oxidizing species $P680^+$ which has a redox potential of +1200 mV, sufficient to oxidize water. Nonetheless, since water splitting involves four electron chemistry and charge separation only involves transfer of one electron, four separate charge separations (turnovers of PSII) are required to drive formation of one molecule of O_2 from two molecules of water. The initial electron donation to generate the P680 from $P680^+$ is therefore provided by a cluster of manganese ions within the oxygen-evolving complex (OEC), which is attached to the lumen side of PSII. Manganese is a transition metal that can exist in a range of oxidation states from +1 to +5 and thus accumulates the positive charges derived from each light-driven turnover of P680. Progressive extraction of electrons from the manganese cluster is driven by the oxidation of P680 within PSII by light and is known as the S-state cycle. After the fourth turnover of P680, sufficient positive charge is built up in the manganese cluster to permit the splitting of water into electrons, which regenerate the original state of the manganese cluster, protons, which are released into the lumen and contribute to the proton gradient used for ATP synthesis, and the by-product O_2. Thus charge separation at P680 provides the thermodynamic driving force, whereas the manganese cluster acts as a catalyst for the water-splitting reaction.

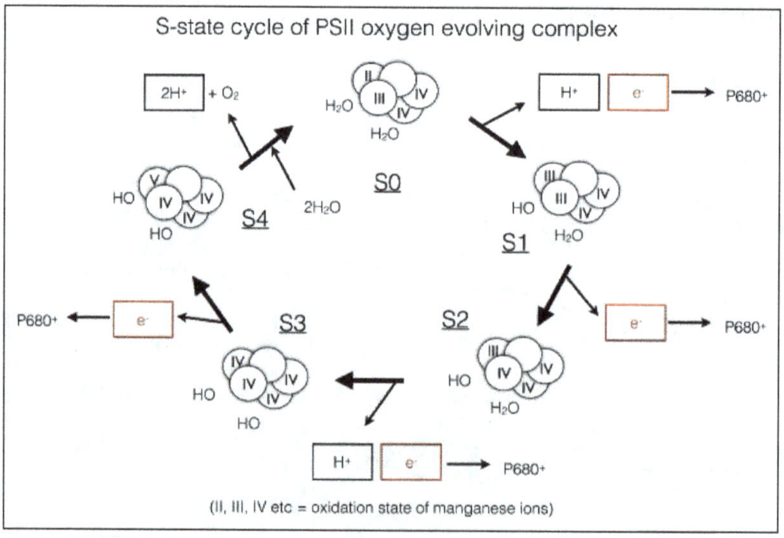

S-state cycle of water oxidation by the manganese cluster (shown as circles with roman numerals representing the manganese ion oxidation states) within the PSII oxygen-evolving complex.

Progressive extraction of electrons from the manganese cluster is driven by the oxidation of P680 within PSII by light. Each of the electrons given up by the cluster is eventually repaid at the S_4 to S_0 transition when molecular oxygen (O_2) is formed. The protons extracted from water during the process are deposited into the lumen and contribute to the protonmotive force.

The electrons yielded by P680* following charge separation are not passed directly to plastoquinone, but rather via another acceptor called pheophytin, a porphyrin molecule lacking the central magnesium ion as in chlorophyll. Plastoquinone reduction to plastoquinol requires two electrons and thus two molecules of plastoquinol are formed per O_2 molecule evolved by PSII. Two protons are also taken up upon formation of plastoquinol and these are derived from the stroma. PSII is found within the thylakoid membrane of plants as a dimeric RC complex surrounded by a peripheral antenna of six minor monomeric antenna LHC complexes and two to eight trimeric LHC complexes, which together form a PSII–LHCII supercomplex.

Photosystem I

PSI is a light-driven plastocyanin–ferredoxin oxidoreductase. In PSI, the special pair of chlorophylls are known as P700 due to their 700 nm absorption peak in the red part of the spectrum. P700* is an extremely strong reductant that is able to reduce ferredoxin which has a redox potential of −450 mV (and is thus is, in principle, a poor electron acceptor). Reduced ferredoxin is then used to generate NADPH for the Calvin–Benson cycle at a separate complex known as FNR. The electron from P700* is donated via another chlorophyll molecule and a bound quinone to a series of iron–sulfur clusters at the stromal side of the complex, whereupon the electron is donated to ferredoxin. The P700 species is regenerated form $P700^+$via donation of an electron from the soluble electron carrier protein plastocyanin.

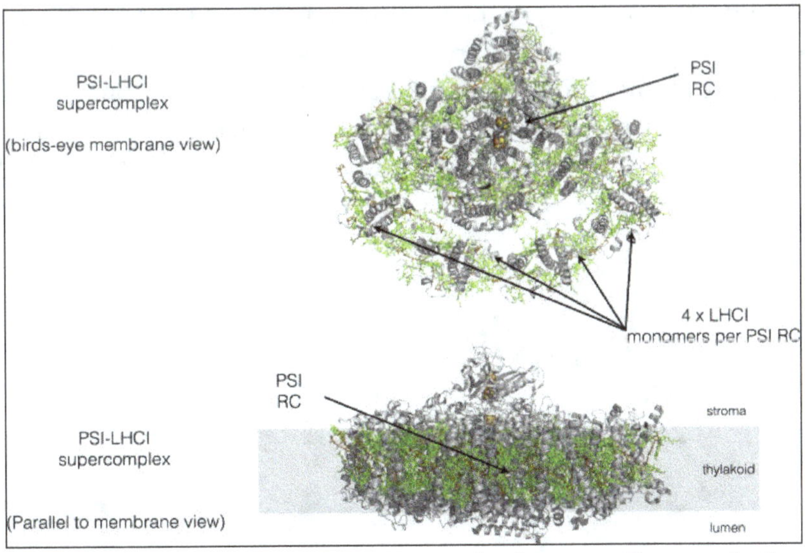

Basic structure of the PSI–LHCI supercomplex from pea. The organization of PSI and its light-harvesting antenna. Protein is shown in grey, with chlorophylls in green and carotenoids in orange.

PSI is found within the thylakoid membrane as a monomeric RC surrounded on one side by four LHC complexes known as LHCI. The PSI–LHCI supercomplex is found mainly in the unstacked regions of the thylakoid membrane.

Other Electron Transfer Chain Components

Plastoquinone/Plastoquinol

Plastoquinone is a small lipophilic electron carrier molecule that resides within the thylakoid membrane and carries two electrons and two protons from PSII to the cytb_6f complex. It has a very similar structure to that of the molecule ubiquinone (coenzyme Q_{10}) in the mitochondrial inner membrane.

Cytochrome b_6f Complex

The cytb_6f complex is a plastoquinol–plastocyanin oxidoreductase and possess a similar structure to that of the cytochrome bc_1 complex (complex III) in mitochondria. As with Complex III, cytb_6f exists as a dimer in the membrane and carries out both the oxidation and reduction of quinones via the so-called Q-cycle. The Q-cycle involves oxidation of one plastoquinol molecule at the Qp site of the complex, both protons from this molecule are deposited in the lumen and contribute to the proton gradient for ATP synthesis. The two electrons, however, have different fates. The first is transferred via an iron–sulfur cluster and a haem cofactor to the soluble electron carrier plastocyanin. The second electron derived from plastoquinol is passed via two separate haem cofactors to another molecule of plastoquinone bound to a separate site (Qn) on the complex, thus reducing it to a semiquinone. When a second plastoquinol molecule is oxidized at Qp, a second molecule of plastocyanin is reduced and two further protons are deposited in the lumen. The second electron reduces the semiquinone at the Qn site which, concomitant with uptake of two protons from the stroma, causes its reduction to plastoquinol. Thus for each pair of plastoquinol molecules oxidized by the complex, one is regenerated, yet all four protons are deposited into the lumen. The Q-cycle thus doubles the number of protons transferred from the stroma to the lumen per plastoquinol molecule oxidized.

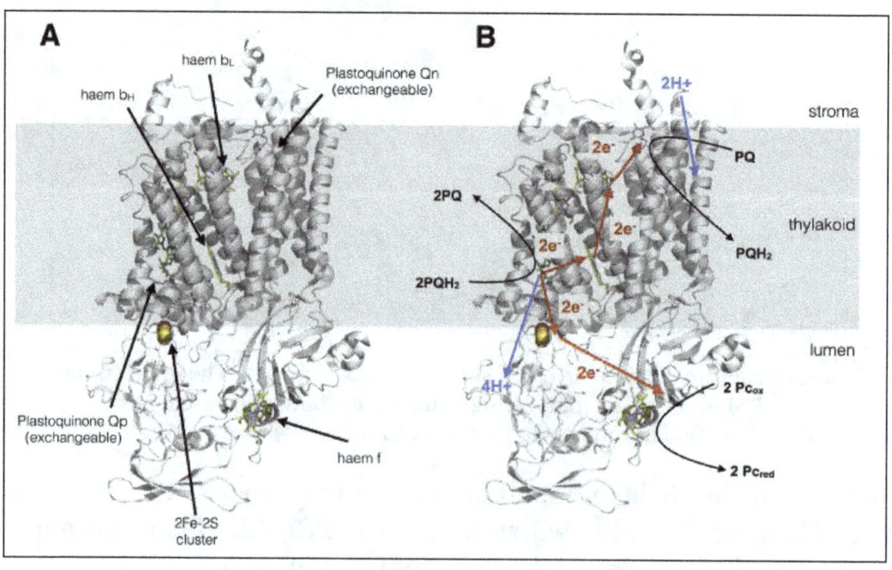

Cytochrome b6f complex: (A) Structure drawn from PDB code 1Q90. (B) The protonmotive Q-cycle showing how electrons from plastoquinol are passed to both plastocyanin and plastoquinone, doubling the protons deposited in the lumen for every plastoquinol molecule oxidized by the complex.

Plastocyanin

Plastocyanin is a small soluble electron carrier protein that resides in the thylakoid lumen. The active site of the plastocyanin protein binds a copper ion, which cycles between the Cu^{2+} and Cu^{+} oxidation states following its oxidation by PSI and reduction by cytb_6f respectively.

Ferredoxin

Ferredoxin is a small soluble electron carrier protein that resides in the chloroplast stroma. The active site of the ferredoxin protein binds an iron–sulfur cluster, which cycles between the Fe^{2+} and Fe^{3+} oxidation states following its reduction by PSI and oxidation by the FNR complex respectively.

Ferredoxin–NADP$^+$ Reductase

The FNR complex is found in both soluble and thylakoid membrane-bound forms. The complex binds a flavin–adenine dinucleotide (FAD) cofactor at its active site, which accepts two electrons from two molecules of ferredoxin before using them reduce NADP$^+$ to NADPH.

ATP Synthase

The ATP synthase enzyme is responsible for making ATP from ADP and P_i; this endergonic reaction is powered by the energy contained within the protonmotive force. According to the structure, 4.67 H^+ are required for every ATP molecule synthesized by the chloroplast ATP synthase. The enzyme is a rotary motor which contains two domains: the membrane-spanning F_O portion which conducts protons from the lumen to the stroma, and the F_1 catalytic domain that couples this exergonic proton movement to ATP synthesis.

Membrane Stacking and the Regulation of Photosynthesis

Within the thylakoid membrane, PSII–LHCII supercomplexes are packed together into domains known as the grana, which associate with one another to form grana stacks. PSI and ATP synthase are excluded from these stacked PSII–LHCII regions by steric constraints and thus PSII and PSI are segregated in the thylakoid membrane between the stacked and unstacked regions. The cytb_6f complex, in contrast, is evenly distributed throughout the grana and stromal lamellae. The evolutionary advantage of membrane stacking is believed to be a higher efficiency of electron transport by preventing the fast energy trap PSI from 'stealing' excitation energy from the slower trap PSII, a phenomenon known as spillover. Another possible advantage of membrane stacking in thylakoids may be the segregation of the linear and cyclic electron transfer pathways, which might otherwise compete to reduce plastoquinone. In this view, PSII, cytb_6f and a sub-fraction of PSI closest to the grana is involved in linear flow, whereas PSI and cytb_6f in the stromal lamellae participates in cyclic flow. The cyclic electron transfer pathway recycles electrons from ferredoxin back to plastoquinone and thus allows protonmotive force generation (and ATP synthesis) without net NADPH production. Cyclic electron transfer thereby provides the additional ATP required for the Calvin–Benson cycle.

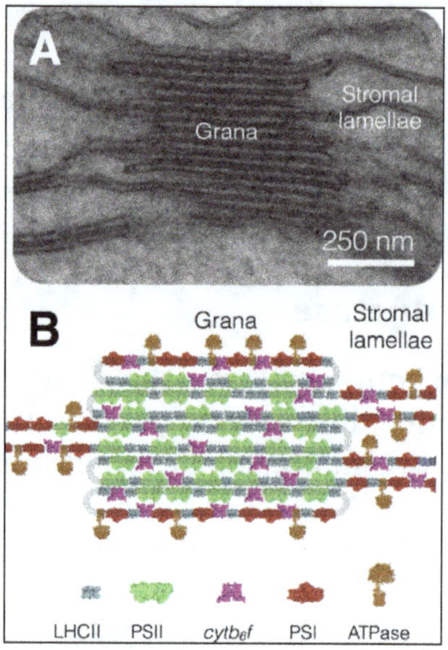

Lateral heterogeneity in thylakoid membrane organization: (A) Electron micrograph of the thylakoid membrane showing stacked grana and unstacked stromal lamellae regions. (B) Model showing the distribution of the major complexes of photosynthetic electron and proton transfer between the stacked grana and unstacked stromal lamellae regions.

Dark Reactions: The Calvin–Benson Cycle

CO_2 is fixed into carbohydrate via the Calvin–Benson cycle in plants, which consumes the ATP and NADPH produced during the light reactions and thus in turn regenerates ADP, P_i and $NADP^+$. In the first step of the Calvin–Benson cycle, CO_2 is combined with a 5-carbon (5C) sugar, ribulose 1,5-bisphosphate in a reaction catalysed by the enzyme ribulose-1,5-bisphosphate carboxylase/oxygenase (Rubisco). The reaction forms an unstable 6C intermediate that immediately splits into two molecules of 3-phosphoglycerate. 3-Phosphoglycerate is first phosphorylated by 3-phosphoglycerate kinase using ATP to form 1,3-bisphosphoglycerate. 1,3-Bisphosphoglycerate is then reduced by glyceraldehyde 3-phosphate dehydrogenase using NADPH to form glyceraldehyde 3-phosphate (GAP, a triose or 3C sugar) in reactions, which are the reverse of glycolysis. For every three CO_2 molecules initially combined with ribulose 1,5-bisphopshate, six molecules of GAP are produced by the subsequent steps. However only one of these six molecules can be considered as a product of the Calvin–Benson cycle since the remaining five are required to regenerate ribulose 1,5-bisphosphate in a complex series of reactions that also require ATP. The one molecule of GAP that is produced for each turn of the cycle can be quickly converted by a range of metabolic pathways into amino acids, lipids or sugars such as glucose. Glucose in turn may be stored as the polymer starch as large granules within chloroplasts.

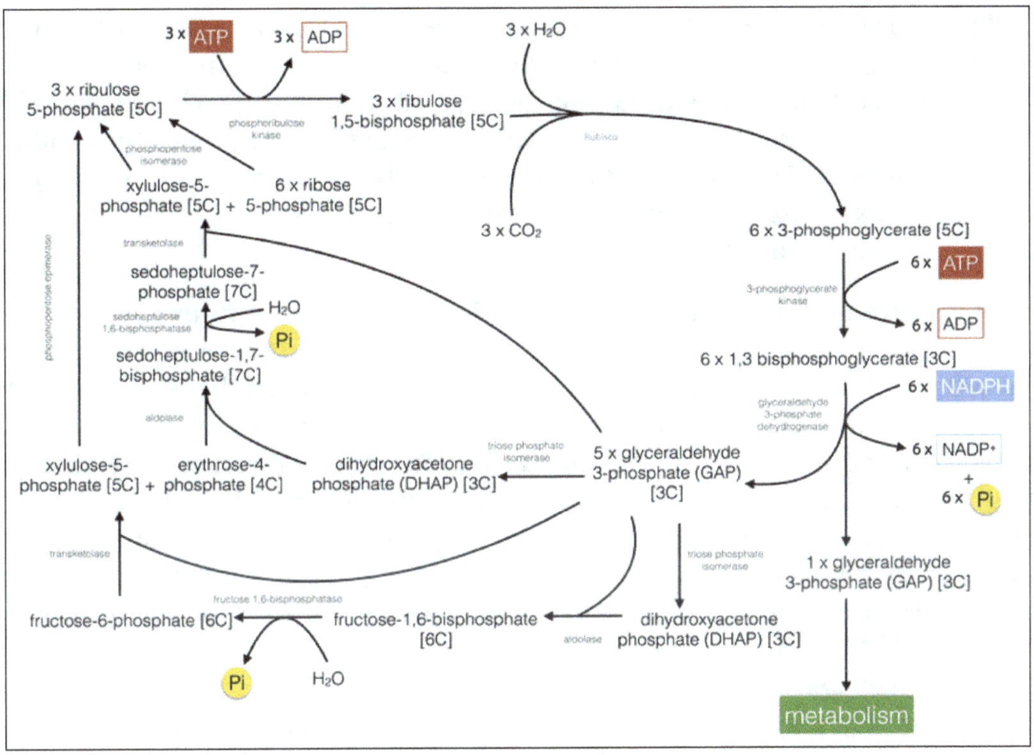

The Calvin–Benson cycle: The biochemical pathway for the fixation of CO_2 into carbohydrate in plants.

A complex biochemical 'dance' is then involved in the regeneration of three ribulose 1,5-bisphosphate (5C) from the remaining five GAP (3C) molecules. The regeneration begins with the conversion of two molecules of GAP into dihydroxyacetone phosphate (DHAP) by triose phosphate isomerase; one of the DHAP molecules is the combined with another GAP molecule to make fructose 1,6-bisphosphate (6C) by aldolase. The fructose 1,6-bisphosphate is then dephosphorylated by fructose-1,6-bisphosphatase to yield fructose 6-phosphate (6C) and releasing P_i. Two carbons are then removed from fructose 6-phosphate by transketolase, generating erythrose 4-phosphate (4C); the two carbons are transferred to another molecule of GAP generating xylulose 5-phosphate (5C). Another DHAP molecule, formed from GAP by triose phosphate isomerase is then combined with the erythrose 4-phosphate by aldolase to form sedoheptulose 1,7-bisphosphate (7C). Sedoheptulose 1,7-bisphosphate is then dephosphorylated to sedoheptulose 7-phosphate (7C) by sedoheptulose-1,7-bisphosphatase releasing P_i. Sedoheptulose 7-phosphate has two carbons removed by transketolase to produce ribose 5-phosphate (5C) and the two carbons are transferred to another GAP molecule producing another xylulose 5-phosphate (5C). Ribose 5-phosphate and the two molecules of xylulose 5-phosphate (5C) are then converted by phosphopentose isomerase to three molecules of ribulose 5-phosphate (5C). The three ribulose 5-phosphate molecules are then phosphorylated using three ATP by phosphoribulokinase to regenerate three ribulose 1,5-bisphosphate (5C).

Overall the synthesis of 1 mol of GAP requires 9 mol of ATP and 6 mol of NADPH, a required ratio of 1.5 ATP/NADPH. Linear electron transfer is generally thought to supply ATP/NADPH in a ratio of 1.28 (assuming an H^+/ATP ratio of 4.67) with the shortfall of ATP believed to be provided by cyclic electron transfer reactions. Since the product of the Calvin cycle is GAP (a 3C sugar) the pathway is often referred to as C_3 photosynthesis and plants that utilize it are called C_3 plants and include many of the world's major crops such as rice, wheat and potato.

Many of the enzymes involved in the Calvin–Benson cycle (e.g. transketolase, glyceraldehyde-3-phosphate dehydrogenase and aldolase) are also involved in the glycolysis pathway of carbohydrate degradation and their activity must therefore be carefully regulated to avoid futile cycling when light is present, i.e. the unwanted degradation of carbohydrate. The regulation of the Calvin–Benson cycle enzymes is achieved by the activity of the light reactions, which modify the environment of the dark reactions (i.e. the stroma). Proton gradient formation across the thylakoid membrane during the light reactions increases the pH and also increases the Mg^{2+} concentration in the stroma (as Mg^{2+} flows out of the lumen as H^+ flows in to compensate for the influx of positive charges). In addition, by reducing ferredoxin and $NADP^+$, PSI changes the redox state of the stroma, which is sensed by the regulatory protein thioredoxin. Thioredoxin, pH and Mg^{2+} concentration play a key role in regulating the activity of the Calvin–Benson cycle enzymes, ensuring the activity of the light and dark reactions is closely co-ordinated.

Rubisco

It is noteworthy that, despite the complexity of the dark reactions, the carbon fixation step itself (i.e. the incorporation of CO_2 into carbohydrate) is carried out by a single enzyme, Rubisco. Rubisco is a large multisubunit soluble protein complex found in the chloroplast stroma. The complex consists of eight large (56 kDa) subunits, which contain both catalytic and regulatory domains, and eight small subunits (14 kDa), which enhance the catalytic function of the L subunits. The carboxylation reaction carried out by Rubisco is highly exergonic ($\Delta G^{o}=-51.9\ kJ{\cdot}mol^{-1}$), yet kinetically very slow (just 3 s^{-1}) and begins with the protonation of ribulose 1,5-bisphosphate to form an enediolate intermediate which can be combined with CO_2 to form an unstable 6C intermediate that is quickly hydrolysed to yield two 3C 3-phosphoglycerate molecules. The active site in the Rubisco enzyme contains a key lysine residue, which reacts with another (non-substrate) molecule of CO_2 to form a carbamate anion that is then able to bind Mg^{2+}. The Mg^{2+} in the active site is essential for the catalytic function of Rubisco, playing a key role in binding ribulose 1,5-bisphosphate and activating it such that it readily reacts with CO_2. Rubisco activity is co-ordinated with that of the light reactions since carbamate formation requires both high Mg^{2+} concentration and alkaline conditions, which are provided by the light-driven changes in the stromal environment.

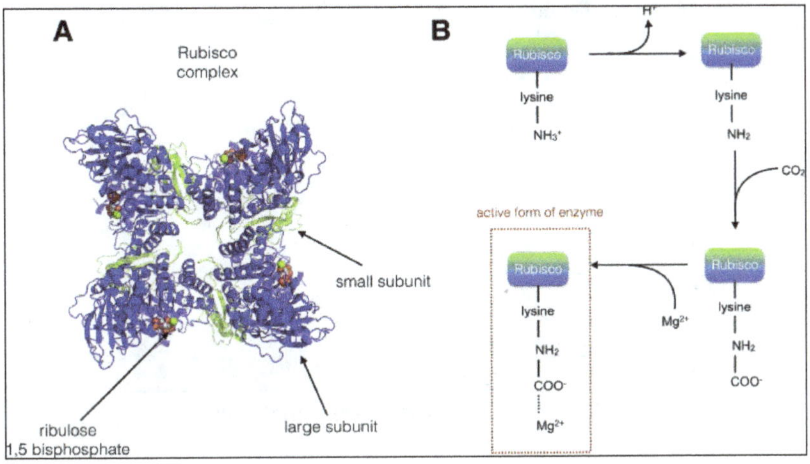

Rubisco: (A) Structure of the Rubisco enzyme (the large subunits are shown in blue and the small subunits in green); four of each type of subunit are visible in the image. Drawn from PDB code 1RXO. (B) Activation of the lysine residue within the active site of Rubisco occurs via elevated stromal pH and Mg^{2+} concentration as a result of the activity of the light reactions.

In addition to carboxylation, Rubisco also catalyses a competitive oxygenation reaction, known as photorespiration, that results in the combination of ribulose 1,5-bisphosphate with O_2 rather than CO_2. In the oxygenation reaction, one rather than two molecules of 3-phosphoglycerate and one molecule of a 2C sugar known as phosphoglycolate are produced by Rubisco. The phosphoglycolate must be converted in a series of reactions that regenerate one molecule of 3-phosphoglycerate and one molecule of CO_2. These reactions consume additional ATP and thus result in an energy loss to the plant. Although the oxygenation reaction of Rubisco is much less favourable than the carboxylation reaction, the relatively high concentration of O_2 in the leaf (250 μM) compared with CO_2 (10 μM) means that a significant amount of photorespiration is always occurring. Under normal conditions, the ratio of carboxylation to oxygenation is between 3:1 and 4:1. However, this ratio can be decreased with increasing temperature due to decreased CO_2 concentration in the leaf, a decrease in the affinity of Rubisco for CO_2 compared with O_2 and an increase in the maximum rate of the oxygenation reaction compared with the carboxylation reaction. The inefficiencies of the Rubisco enzyme mean that plants must produce it in very large amounts (~30–50% of total soluble protein in a spinach leaf) to achieve the maximal photosynthetic rate.

CO_2 Concentrating Mechanisms

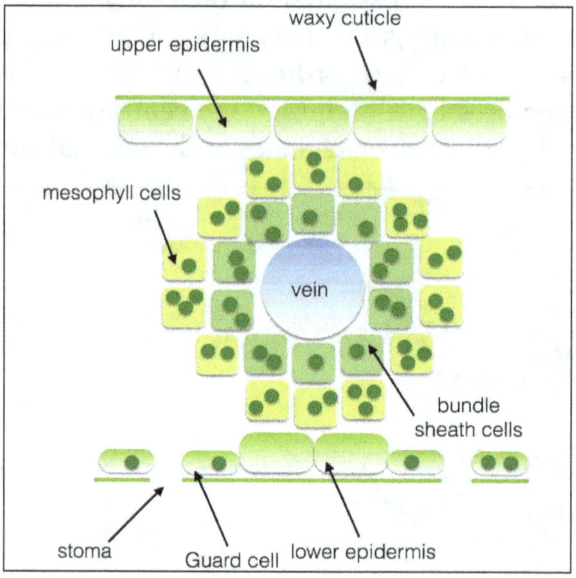

C_4 plant leaf showing Kranz anatomy.

To counter photorespiration, plants, algae and cyanobacteria have evolved different CO_2-concentrating mechanisms CCMs that aim to increase the concentration of CO_2 relative to O_2 in the vicinity of Rubisco. One such CCM is C_4 photosynthesis that is found in plants such as maize, sugar cane and savanna grasses. C_4 plants show a specialized leaf anatomy: Kranz anatomy. Kranz, German for wreath, refers to a bundle sheath of cells that surrounds the central vein within the leaf, which in turn are surrounded by the mesophyll cells. The mesophyll cells in such leaves are rich in the enzyme phosphoenolpyruvate (PEP) carboxylase, which fixes CO_2 into a 4C carboxylic acid: oxaloaceatate. The oxaloacetate formed by the mesophyll cells is reduced using NADPH to malate, another 4C acid: Malate. The malate is then exported from the mesophyll cells to the bundle sheath cells, where it is decarboxylated to pyruvate thus regenerating NADPH and CO_2. The CO_2 is then utilized by Rubisco in the Calvin cycle. The pyruvate is in turn returned to the mesophyll cells where it is phosphorylated using ATP to reform PEP. The advantage of C_4 photosynthesis is that CO_2 accumulates at a very high concentration in the bundle sheath cells that is then sufficient to allow Rubisco to operate efficiently.

Plants growing in hot, bright and dry conditions inevitably have to have their stomata closed for large parts of the day to avoid excessive water loss and wilting. The net result is that the internal CO_2 concentration in the leaf is very low, meaning that C_3 photosynthesis is not possible. To counter this limitation, another CCM is found in succulent plants such as cacti. The Crassulaceae fix CO_2 into malate during the day via PEP carboxylase, store it within the vacuole of the plant cell at night and then release it within their tissues by day to be fixed via normal C_3 photosynthesis. This is termed crassulacean acid metabolism (CAM).

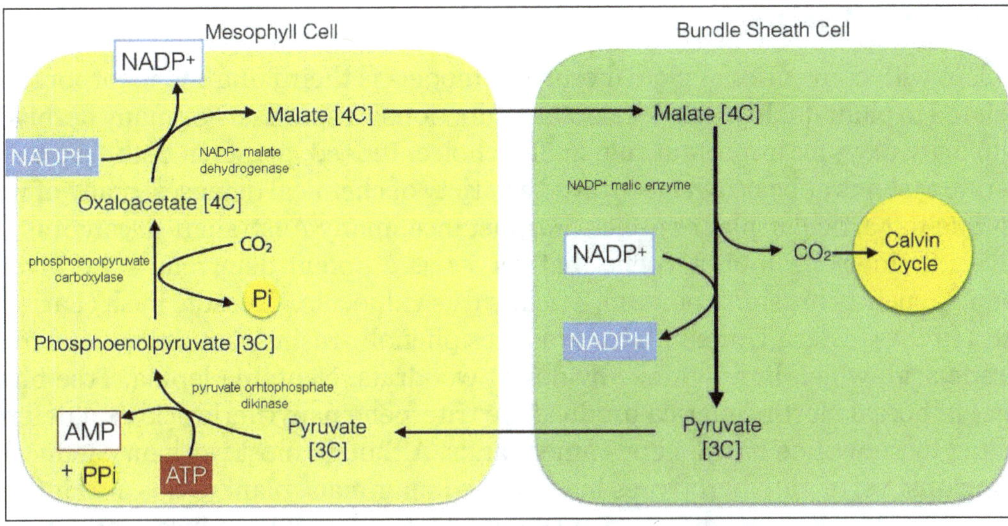

The C_4 pathway ($NADP^+$–malic enzyme type) for fixation of CO_2.

Troublesome Toxins

By definition herbivores depend on plants to survive. The need to obtain suitable food in sufficient amounts drives innumerable herbivore behaviors; for example, movement decisions often are related to the distribution and abundance of plant resources. By the same token, herbivores can exert strong effects on plant growth, survival, and population size by virtue of their feeding habits. Plant demographic effects are especially severe during cyclical peaks or irruptions in herbivore populations. Moreover, the ecological effects of herbivores can extend beyond populations. Differential foraging among species can affect outcomes of competition, facilitate invasion of extant communities, and alter patterns of plant succession, diversity, and dominance.

Conventional Modeling Approaches

When focusing on optimal diet choice by herbivores, ecologists traditionally have relied on linear programming or linear dynamic programming methods. Given a choice of two or more non-equivalent food types, these methods solve for optimal diet composition subject to constraints imposed by daily energy requirements, feeding time, digestive capacity, or nutrient requirements. Linear programming appears to provide reasonable predictions of diet composition for many species. However, it does not address population-level dynamics of herbivores and plants.

Consumer-resource interactions at the population level can be modeled using equations that relate the rate of resource intake by a consumer to resource abundance. These so-called "functional-response" models link herbivore behavior and plant characteristics to population- and community-level consequences. In these models, upper limits to rates of consumption by herbivores are determined, either implicitly or analytically, by combining mechanical factors such as bite size and rate with plant quantity.

Ignore Plant Toxins

A problem with conventional plant-herbivore models is their failure to incorporate factors related to plant quality into decelerating functional responses. For many herbivores, plant toxicity plays an important role in diet choice. Indeed, plants in both tropical and temperate systems appear to have evolved a variety of chemical defenses, many of which are unique to particular plant species. For instance, many Australian Eucalyptus trees produce 1,8-cineole, a monoterpene that serves as a potent deterrent to herbivorous marsupials such as brushtail possum, Trichosurus vulpecula. Creosote bush (Larrea tridentata) in the western United States produces phenolic resins containing nordihydroguaiaretic acid, which limits intake by desert woodrats, Neotoma lepida. Tree birches (Betula) in boreal North America produce the triterpene papyriferic acid as a deterrent to feeding by snowshoe hares, Lepus americanus. Although most work on chemical defenses against vertebrate herbivores has involved mammals, plant toxins also influence herbivorous birds. For instance, aspen (Populus tremuloides) produces coniferyl benzoate, a phenylpropanoid ester that inhibits feeding by ruffed grouse, Bonasa umbellus.

In addition to interspecific differences, production of toxins varies ontogenetically within plants, and among individuals and populations within species. Intraspecific variation in chemical defense often contains strong genetic components. When combined with spatial variation in environmental conditions and herbivory, substantial geographic variation in defense can occur within species. Ontogenetic variation in defensive responses of many plants is shaped by constraints on resource allocation and sensitivity to fitness consequences of herbivory. For instance, winter browsing of plants by mammals has severe repercussions for fitness during the juvenile stage and was linked to greater defense of juveniles in a review of 37 woody species.

Coping Strategies of Herbivores

One consequence of feeding on plants containing toxins is that rates of ingestion may be limited by an herbivore's ability to avoid toxins or detoxify food rather than to mechanically process food. Not surprisingly, herbivores have developed a host of physiological and behavioral mechanisms to deal with plant secondary metabolites. Physiologically, vertebrates can regulate absorption of plant toxins by gut cells, respond to chemically mediated taste and trigeminal stimulation, and detoxify lipophilic compounds via enzymatic biotransformation. For instance, marsupial folivores oxidize plant terpenes using P450 enzymes, and species with diets high in monoterpenes exhibit greater capacity for biotransformation of toxins than their generalist counterparts. Behaviorally, vertebrates can select plants or plant parts containing low concentrations of a toxin, manage food to leach toxins from plants, self-medicate to ameliorate effects of toxins, and adjust meal duration and intake per meal. An ability to regulate intake of plant secondary metabolites has been reported for several species of vertebrate herbivores. For instance, brushtail possums ate more of the toxin benzoate when the rate at which it could be detoxified by conjugation was increased by adding glycine to the diet. Herbivores also achieve greater intake of nutrients by selecting mixed diets containing foods processed by different detoxification pathways, thereby avoiding saturation of any particular pathway. Regardless of the strategies used by herbivores, costs of detoxification often are high. For desert woodrats subsisting on a diet containing a toxin-rich juniper (Juniperus monosperma), detoxification costs are comparable to energy needed for reproduction. For ruffed grouse feeding on aspen, 10 percent of metabolizable energy is lost each day in biotransformation conjugates; additional losses of energy in the conjugation process and of nitrogen due to excretion of amino acid conjugates elevate the cost further. In the face of such costs, vertebrate herbivores face life-history tradeoffs associated with allocation of resources to growth and reproduction.

Needed: A Toxin-determined Functional Response

In light of the widespread nature of plant toxins and their influence on herbivores, a new approach to linking herbivores to their food base is needed if we are to understand implications for herbivore populations and plant communities. Traditional functional responses for vertebrate herbivores have not considered the role of plant toxins. At least three analytical modifications should be considered when incorporating the effects of toxins on plant-herbivore dynamics. Toxin-mediated functional responses should be explicitly accounts for the negative effects of plant toxins on herbivore growth; permitherbivores to regulate intake of toxins; and allow for intake of multiple plants that are detoxified with independent pathways. Recently, progress has been made in the first two areas. Specifically, a conventional functional response has been modified to take the form of C(N):

$$C(N) = f(N)\left(1 - \frac{f(N)}{4G}\right)$$

where

$$f(N) = \frac{e\sigma N}{1 + he\sigma N}$$

The term f(N) is the traditional Holling Type 2 functional response in which N is plant biomass, e is resource encounter rate, h is handling time for each plant, and σ is the fraction of encountered food items that are ingested, thereby allowing herbivores to regulate their intake. The second factor in C(N) explicitly accounts for the negative effect of toxins. The parameter G stands for the ratio M/T, where M is the maximum amount of toxicant per unit time that the herbivore can tolerate and T is the amount of toxicant per unit of plant biomass. The factor 4 simplifies the peak value of C(N) as a function of N. In the limit that $1/G << 1$, the effect of the toxicant can be viewed as purely a slowdown in feeding rate. For example, in that limit C(N) above is an approximation of:

$$C(N) \cong \frac{e\sigma N}{1 + he\sigma N + \frac{e\sigma N}{4G}}$$

So in equation above the presence of toxin simply results in an effective increase in the handling time that is proportional to 1/G.

Related functional response models have been formulated to examine how plant growth is limited when the presence of one resource interferes with another resource or is toxic. In other models, additive effects of nutrient limitation of plant growth have been incorporated:

$$C = \frac{C_{max} N_1}{K_1 + N_1 + K_2 \frac{N_1}{N_2}}, \text{where}$$

N_1 and N_2 represent the concentrations of nutrients limiting plant growth, C_{max} is the maximum possible rate of nutrient-limited plant growth, and K_1 and K_2 are constants reflecting the stoichiometry of the two nutrients in the plant. Note that in the limiting case, as N_1/N_2 approaches zero, equation above reduces to the traditional Michaelis-Menten equation for N_1, with K_1 as a half-saturation coefficient. More importantly from our perspective, increasing availability of the co-limiting nutrient, N_2, causes growth rate to increase in equation $C = \frac{C_{max} N_1}{K_1 + N_1 + K_2 \frac{N_1}{N_2}}$, whereas in equation $C(N) \cong \frac{e\sigma N}{1 + he\sigma N + \frac{e\sigma N}{4G}}$ an increasing concentration of toxin in food relative to the rate of toxin ingestion the herbivore can tolerate, 1/G, decreases growth rate. In both equations, this change in C is due to a change in size of the third term in the denominator.

The toxin-determined functional response differs from plant-nutrient models because toxins can do more than reduce feeding rate. Specifically, when 1/G is large (and hence each gram of plant is quite toxic to herbivores), the functional response can represent a more serious deterioration of the herbivore's ability to feed or survive. Analysis has demonstrated the critical importance of σ to herbivore dynamics; in the presence of a toxin, selection should act strongly to regulate intake below the herbivore's detoxification threshold.

Because most vertebrate herbivores are generalists, a toxin-mediated functional response should be able to consider multiple plant species. Fortunately, the single species framework extends directly to multiple species. Recent analysis of a multi-species model with independent pathways for detoxification yielded predictions that matched remarkably well with empirical studies of moose (Alces alces) and snowshoe hares feeding on woody plants. Under conditions of the multi-species model, herbivores switch feeding on plant species to avoid saturating detoxification systems. One consequence of toxin-induced switching is that herbivores are predicted to spend a disproportionate amount of time foraging on less abundant plant species, resulting in depensatory mortality that can limit invasion by more palatable species. Consistent with this prediction, disproportionate foraging on rare but more palatable food items has been observed in experimental studies with foods that vary in quality. Another likely consequence of toxin-determined herbivory is a shift in plant species composition to communities dominated by more toxic plants, as observed in taiga, southern boreal forest, and temperate grasslands. Toxin-determined foraging also may play a role in population cycles of herbivores.

Evidence for the importance of plant toxins as determinants of herbivore functional response is indisputable. Recent modeling efforts implicate toxins as potentially key drivers of change in plant communities and herbivore populations. Future models should consider the role of resource patchiness and tri-trophic interactions on plant communities. For instance, adaptive foraging by herbivores is hypothesized to have important effects on ecosystem processes such as nutrient cycling rates, and predators may alter herbivore effects by changing their density or behavior. How do tradeoffs from toxin-induced resource patchiness and risk of predation influence ecosystem properties? From the perspective of evolutionary ecology, models of tradeoffs in plant growth and defense as well as spatio-temporal variation in selection for toxin production may afford greater insight into genetic diversity and geographic structuring of plant populations. At the very least, ecologists conducting work in the future should address explicitly the importance of plant toxins as potential agents of change for plant and herbivore communities.

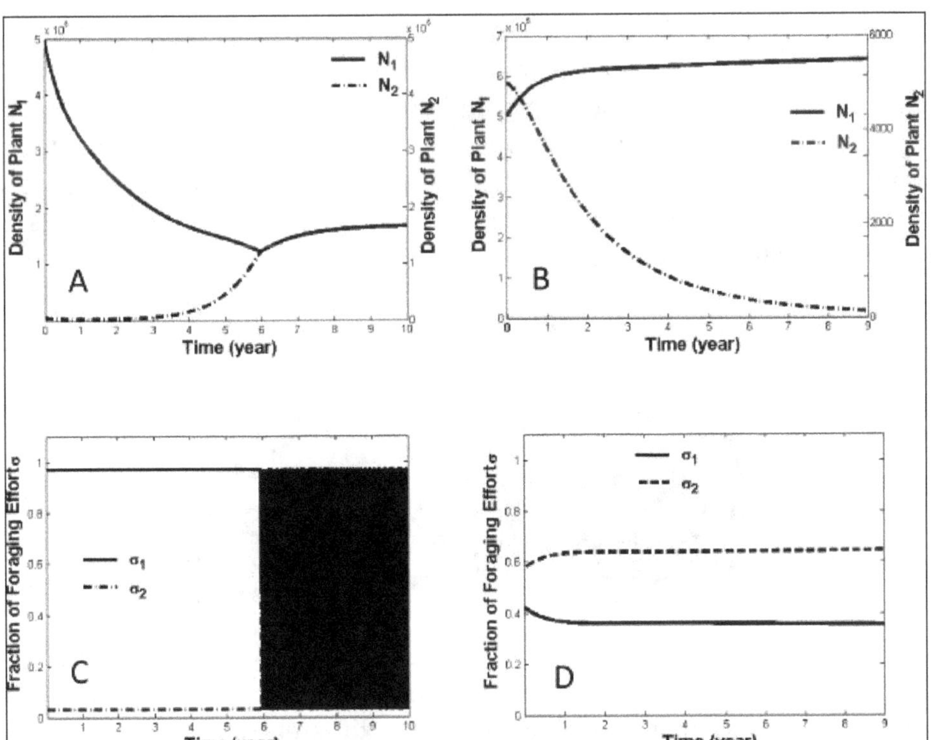

Dynamics of two-species plant communities subjected to a population of herbivores that feed optimally. N_1 and N_2 refer to biomasses of edible plants. The functional response used to generate the plots is the one shown in equation $C(N) = f(N)\left(1 - \frac{f(N)}{4G}\right)$ of the text, modified to permit two plant species. When toxins are not incorporated into the functional response (panel A), the plant species coexist. Herbivores feed exclusively on plant species 1 initially, i.e., $\sigma_1 = 1$, $\sigma_2 = 0$ (panel C), since the starting density of species 1 is higher. Once the density of species 2 exceeds that of species 1, the consumption constants switch to $\sigma_1 = 0$, $\sigma_2 = 1$, and the switches continue to occur (the switches occur so rapidly that it appears as a black area in panel C). When toxins are incorporated into the functional response and the resident plant species (species 1 in panel B) is more toxic than a prospective invading species (species 2 in panel B), simulation results demonstrate that the less toxic plant fails to establish. The failure is tied to the adaptive foraging behavior of the herbivore, resulting in a disproportionate fraction of its effort being expended on the less abundant (but less toxic) species 2 (panel D). Parameter values: $c_{12} = 0.9$, $c_{21} = 0.9$, $r_1 = r_2 = 0.007$, $K_1 = K_2 = 7*10^5$, $B_1 = B_2 = 3.4*10^{-5}$, $e_1 = e_2 = 0.0007$, $h_1 = h_2 = 0.008$, mp = 0.0013. For simulations of plants containing toxins (panels B and D): $G_1 = 35$, $G_2 = 60$, initial density of species 1 = 5×10^5, initial density of species 2 = 5×10^3.

Pesticide Chemistry

Pesticide Biochemistry deals with the mode of action of plant protection agents such as insecticides, fungicides, herbicides, and similar compounds, including nonlethal pest control agents, biosynthesis of pheromones, hormones, and plant resistance agents. It mainly emphasizes on Biochemistry and Physiology of Comparative toxicity, Mode of action, Pathophysiology, Plant growth regulators, Resistance, Other effects of pesticides on both parasites and hosts.

Pesticides are chemicals that kill pests, and are categorized by the types of pests they kill. For example, insecticides kill insects, herbicides kill weeds, bactericides kill bacteria, fungicides kill fungi, and algicides kill algae. Approximately 90 percent of all pesticides used worldwide are used in agriculture, food storage, or shipping. Because of a growing world population, there is pressure to increase and preserve the food supply by using pesticides and other agricultural chemicals.

Classes of Pesticides

Although the use of pesticides is not new, the types of substances people have used as pesticides have changed over time. The earliest pesticides were inorganic substances such as sulfur, mercury, lead, arsenic, and ash. Some of these inorganic pesticides are still used today. For example, sulfur is still used as a fungicide, copper is used as an algicide, lead and arsenic were used as insecticides until World War II, and chromium, copper, and arsenic have been used as wood preservatives to prevent microorganisms from causing wood decay. Even though many of these substances are effective pesticides, the use of some of these materials has been banned or restricted because of health and environmental concerns. Lead and arsenic are no longer used as insecticides, the use of mercury as a fungicide has been restricted, and the U.S. Environmental Protection Agency (EPA) is phasing out the use of arsenic as a wood preservative.

The modern era of chemical pest control began around the time of World War II, when the synthetic organic chemical industry began to develop. The first synthetic organic pesticides were organochlorine compounds, such as dichlorodiphenyltrichloroethane (DDT). Commercial production of DDT began in 1943. At that time, DDT was considered to be a wonderful invention. It was cheap to produce, very toxic to insects, and much less toxic to mammals. DDT and other organochlorine insecticides were used for many years to control mosquitoes and as a broad-spectrum insecticide against insect pests that damaged food and crops. Unfortunately, scientists learned later that many organochlorine insecticides were persistent in the environment (they did not degrade readily) and were bioaccumulating in birds, humans, and other animals. In 1962 Rachel Carson wrote the book Silent Spring, in which she reported that DDT was causing eggshell thinning in bird eggs and thus was leading to the near extinction of bird species such as peregrine falcons and bald eagles. Today most of the organochlorine pesticides have been banned in the United States by the EPA because of the tendency of these compounds to persist in the environment and bioaccumulate in animals.

Other classes of insecticides include the organophosphates, carbamates, pyrethroids, and biopesticides. These other classes of pesticides are not as persistent in the environment as the organochlorine pesticides. The organophosphate and carbamate pesticides affect the nervous system by disrupting the enzyme that regulates acetylcholine a neurotransmitter. However, carbamate pesticides are less toxic to humans because their interactions with important enzymes are reversible. As a group, the organophosphate and carbamate pesticides are probably the most widely used insecticides, although many are being restricted by the EPA because of their toxicity.

Pyrethroid pesticides were developed as synthetic versions of the naturally occurring pesticide pyrethrin, which is found in chrysanthemums. Most pyrethroids are safer than the organochlorines, organophosphates, and carbamates, although some synthetic pyrethroids are toxic to the nervous system. Pyrethroids have been modified to increase their stability in the environment, and many different pyrethroids are being used today.

Pesticide Residues

Pesticide residues are the materials that remain on plants and food when a crop is treated with a pesticide. The U.S. government establishes safe residue levels, called tolerances or maximum residue levels, for each food commodity. However, the presence of pesticide residues in food has been a public concern. There has also been a concern about pesticide residues in water, air, and soil. In response to this concern, the U.S. Congress passed the Food Quality Protection Act in 1996, which has had an impact on safety standards for pesticides.

Approaches to pest management have changed significantly since the 1950s and will continue to change as scientists learn more about the toxicity and environmental behavior of pesticides. Scientists will continue to develop newer approaches to insect pest management that are considered to be safer than the use of broad-spectrum pesticides. The most effective strategy for controlling pests may be to combine methods in an approach known as integrated pest management (IPM), which emphasizes preventing pest damage. In IPM information about pests and available pest control methods is used to manage pest damage by the most economical means and causing the least possible hazard to people, property, and the environment. Methods for pest management will continue to evolve as scientists conduct research and develop new information.

5
Green Chemistry

The branch of chemistry and chemical engineering that focuses on designing of processes and products for mitigating the use of hazardous substances is termed as green chemistry. A few of its aspects are green chemistry metrics and principles, etc. The topics elaborated in this chapter will help in gaining a better perspective of green chemistry.

Green chemistry, also called sustainable chemistry is an approach to chemistry that endeavours to prevent or reduce pollution. This discipline also strives to improve the yield efficiency of chemical products by modifying how chemicals are designed, manufactured, and used.

Green chemistry dates from 1991, when the U.S. Environmental Protection Agency (EPA) launched the Alternative Synthetic Pathways for Pollution Prevention research program under the auspices of the Pollution Prevention Act of 1990. This program marked a radical departure from previous EPA initiatives in emphasizing the reduction or elimination of the production of hazardous substances, as opposed to managing these chemicals after they were manufactured and released into the environment. This research program later expanded to include the development of greener solvents and safer chemicals. The name green chemistry was officially adopted in 1996.

The goal of the Pollution Prevention Act of 1990 was not simply to regulate the quantity and type of emissions but to place limits on the industry in order to reduce the amount of pollution it generated. American chemist Paul Anastas, one of the principal founders of green chemistry, claimed that by improving how chemicals are synthesized, it might be possible to prevent the production of pollutants.

Atom Economy

Of these principles, "atom economy," originally suggested by American chemist Barry Trost in 1973, became a central concept among green chemistry researchers. Atom economy was designed to overcome the limitations of the traditional concept of "yield," the amount of final products, which was used for calculating the efficiency of chemical reactions. To calculate the yield, chemists traditionally considered only the amount of the main chemical product they intended to produce ("target molecules") and not by-products, which might include environmentally hazardous materials. In contrast, atom economy takes into account all reactants and products and hence provides a more-reliable indicator of whether the reaction produces undesirable by-products—that is, pollutants. Green chemistry has since demonstrated that high-efficiency atom economy is indeed achievable through such processes as hydrogenation, metathesis, and cycloaddition.

The concept of greening chemistry developed in the business and regulatory communities as a natural evolution of pollution prevention initiatives. In our efforts to improve crop protection, commercial products and medicines, we also caused unintended harm to our planet and humans.

By the mid-20th century, some of the long-term negative effects of these advancements could not be ignored. Pollution choked many of the world's waterways and acid rain deteriorated forest health. There were measurable holes in the earth's ozone. Some chemicals in common use were suspected of causing or directly linked to human cancer and other adverse human and environmental health outcomes. Many governments began to regulate the generation and disposal of industrial wastes and emissions. The United States formed the Environmental Protection Agency (EPA) in 1970, which was charged with protecting human and environmental health through setting and enforcing environmental regulations.

Green chemistry takes the EPA's mandate a step further and creates a new reality for chemistry and engineering by asking chemists and engineers to design chemicals, chemical processes and commercial products in a way that, at the very least, avoids the creation of toxics and waste. We are able to develop chemical processes and earth-friendly products that will prevent pollution in the first place. Through the practice of green chemistry, we can create alternatives to hazardous substances. We can design chemical processes that reduce waste and reduce demand on diminishing resources. We can employ processes that use smaller amounts of energy. We can do all of this and still maintain economic growth and opportunities while providing affordable products and services to a growing world population.

Sustainable and green chemistry in very simple terms is just a different way of thinking about how chemistry and chemical engineering can be done. Over the years different principles have been proposed that can be used when thinking about the design, development and implementation of chemical products and processes. These principles enable scientists and engineers to protect and benefit the economy, people and the planet by finding creative and innovative ways to reduce waste, conserve energy, and discover replacements for hazardous substances. It's important to note that the scope of these of green chemistry and engineering principles go beyond concerns over hazards from chemical toxicity and include energy conservation, waste reduction, and life cycle considerations such as the use of more sustainable or renewable feedstocks and designing for end of life or the final disposition of the product.

Green chemistry can also be defined through the use of metrics. While a unified set of metrics has not been established, many ways to quantify greener processes and products have been proposed. These metrics include ones for mass, energy, hazardous substance reduction or elimination, and life cycle environmental impacts.

Green Chemistry for Chemical Synthesis

Green chemistry for chemical synthesis addresses our future challenges in working with chemical processes and products by inventing novel reactions that can maximize the desired products and minimize by-products, designing new synthetic schemes and apparati that can simplify operations in chemical productions and seeking greener solvents that are inherently environmentally and ecologically benign.

Over the past two centuries, fundamental theories and reactivities in chemistry have been soundly established. Such theories and reactivities have provided the foundations for the chemical enterprise that generates critical living needs such as food for the world's population, achieves various medical wonders that save millions of lives and improve people's health, and produces materials essential to the present and future needs of mankind. Just less than two centuries ago, organic compounds were believed to be only accessible through biological processes under the influence of "vital forces". Today, many molecules of great complexity can be synthesized readily. The total syntheses of natural products with extremely high complexity such as vitamin B_{12} and palytoxin in the laboratory are testimonials of achievements comparable to the construction of the great pyramids at the molecular scale. However, despite such enormous achievements, we are facing great challenges in future chemical synthesis. The present state-of-the-art processes for synthesizing chemical products are highly inefficient. The concept of atom economy was created to emphasize the importance of this inefficiency. The E factor provided a quantifiable measure of such inefficiency and showed that, for every kilogram of fine chemical and pharmaceutical products produced, 5–100 times that amount of chemical waste is generated.

Such low efficiency in state-of-the-art organic syntheses presents great challenges in resource conservation and draws environmental and health concerns related to the chemical wastes.

Since its birth over a decade ago the field of Green Chemistry has been specifically designed to meet such challenges in chemical synthesis. To address these challenges, innovative and fundamentally novel chemistry is needed throughout the synthetic processes: feedstocks, reactions, solvents, and separations.

Chemical Feedstocks

Presently, the main feedstock of chemical products comes from nonrenewable petroleum that is being depleted rapidly both for chemical and energy needs. However, nature provides a vast amount of biomass in the renewable forms of carbohydrates, amino acids, and triglcerides to obtain organic products, but a major obstacle to using renewable biomass as feedstock is the need for novel chemistry to transform the large amounts of biomass selectively and efficiently, in its natural state, without extensive functionalization, defunctionalization, or protection.

Reactions

Reactions play the most fundamental role in synthesis. The ideology of Green Chemistry calls for the development of new chemical reactivities and reaction conditions that can potentially provide benefits for chemical syntheses in terms of resource and energy efficiency, product selectivity, operational simplicity, and health and environmental safety.

Atom Economy

Conventionally, attaining the highest yield and product selectivity were the governing factors of chemical synthesis. Little consideration was given to the usage of multiple reagents in stoichiometric quantities, which often were not incorporated into the target molecule and would result in significant side products. However, in a balanced chemical reaction, a simple addition or cyclo addition incorporates all atoms of the starting materials into the final product. Recognizing this fundamental phenomenon, in 1991 Trost presented a set of coherent guiding principles for evaluating the efficiency of specific chemical processes, termed the atom economy, which has subsequently been incorporated into the "Twelve Principles of Green Chemistry" and has altered the way many chemists design and plan their syntheses. Atom economy seeks to maximize the incorporation of the starting materials into the final product of any given reaction. The additional corollary is that, if maximum incorporation cannot be achieved, then ideally the quantities of side products should be minute and environmentally innocuous. There is a fundamental difference in the manner in which a reaction yield and the atom economy yield is calculated.

Definition of the fundamental difference in the manner in which the reaction and the atom economy yields are generated:

$$\text{Reaction yield} = \frac{\text{quantity of product isolated}}{\text{theoretical quantity of product}} \times 100\%$$

$$\text{Atom Economy} = \frac{\text{molecular wt. of desired product}}{\text{molecular weight of all product}} \times 100\%$$

The reaction yield is only concerned with the quantity of the desired product that is isolated, relative to the theoretical quantity of the product. Atom economy takes all used reagents and unwanted side products into account along with the desired product. For example, substitutions and eliminations represent the vast majority of uneconomical classical reactions in which inherent wastes are unavoidable. Simple additions or cycloadditions and rearrangements represent desired modes of reactivities. Reaction Mass Efficiency (RME) and Mass Intensity (MI) are additional concepts to evaluate the efficiency of synthetic reactions to take into account the reaction yield.

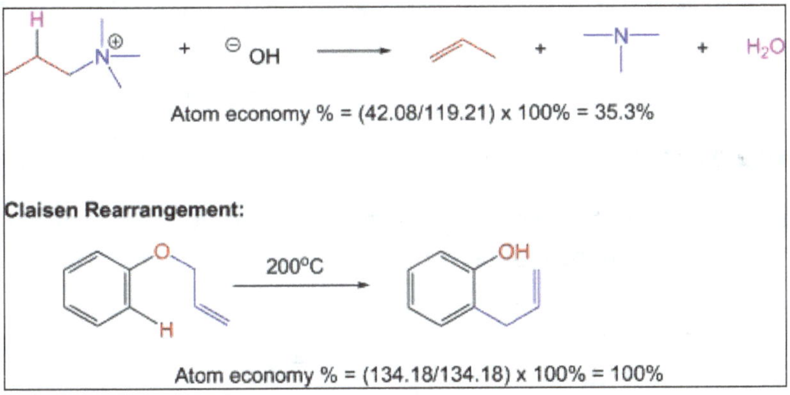

Recently, innovative reactions with such inherent advantages have been developed with the aid of chemical and biological catalysts. Some representative examples are the following:

Isomerizations

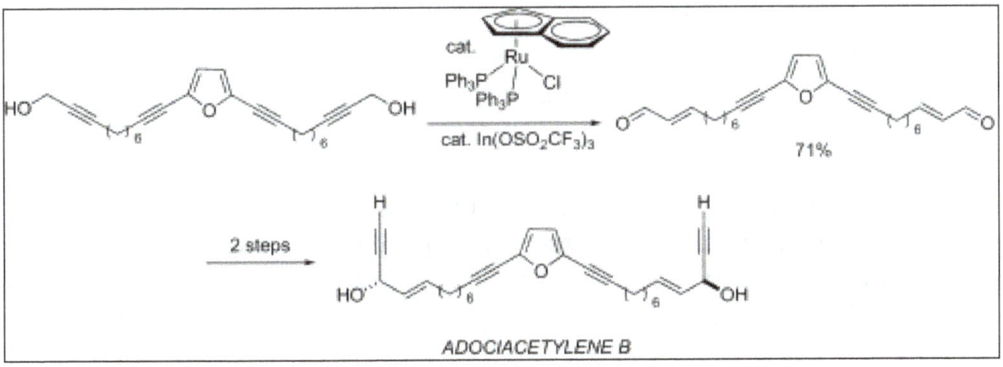

Isomerization of propargyl alcohols into conjugated carbonyl compounds provides an atom-economic means for synthesizing such compounds. By using a ruthenium-catalyzed redox isomerization of propargyl alcohols into enones in lieu of the traditional two-step stoichiometric reduction and oxidation sequence, a catalytic enantioselective total synthesis of adociacetylene B can be realized efficiently. Isomerization of an alkynyl vinylcyclopropene to a fused 5- to 7-ring structure converts classical atom inefficient synthetic strategies to ones of ideal atom economy.

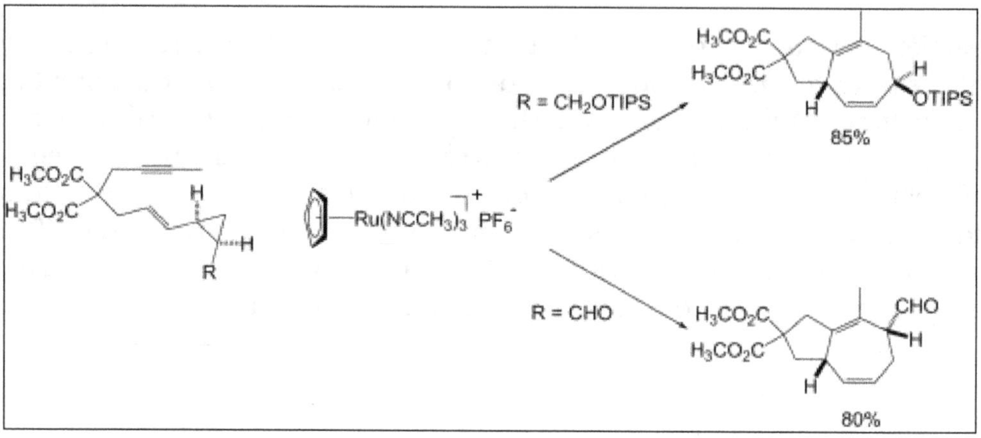

Ring-opening Metathesis Polymerization

The living ring-opening metathesis polymerization developed by Grubbs and coworkers represents another type of isomerization and has been used to make a variety of materials, such as those used in dentistry.

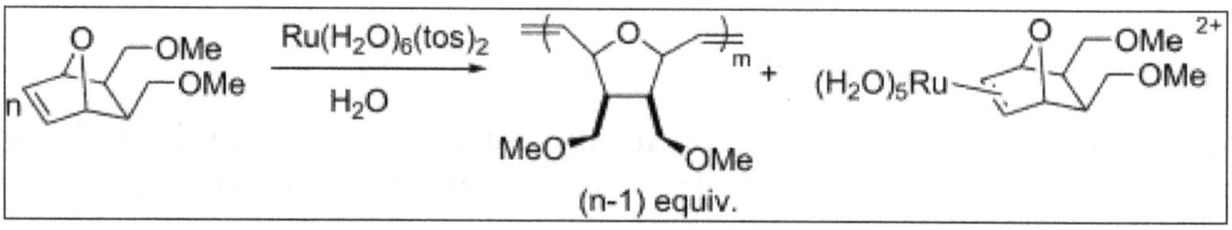

Addition Reactions

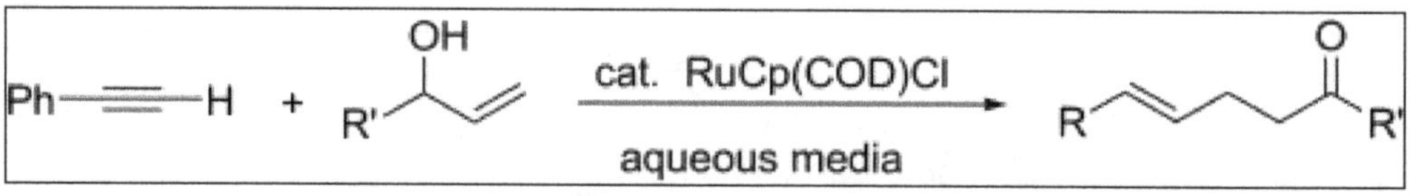

The processes of adding allyl alcohol to alkynes to form γ,δ-unstaturated ketones and aldehydes in aqueous media were developed by Trost et al. and Dixneuf et al., respectively. Another elegant example was recently reported by Krische and coworkers, in which primary alcohols were added stereoselectively to alkenes, which provides an atom-economic version of the classical reaction where a Grignard reagent is added to an aldehyde.

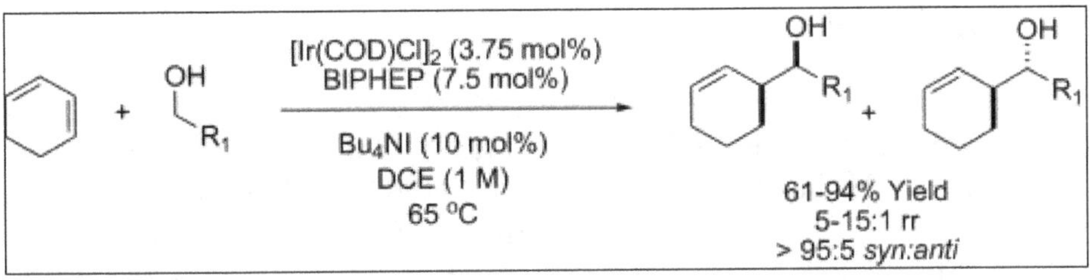

Direct Conversion of C–H Bonds

Direct transformation of the C–H bonds of organic molecules into desired structures without extra chemical transformations represents another class of major desirable reactions. In nature, a variety of organic compounds can be oxidized easily by molecular oxygen or other oxygen donors in the cells of bacteria, fungi, plants, insects, fish, and mammals. It is worth noting the important advances in biomimetic approaches to such oxidations. Hydroxylation of linear alkanes or methane to generate terminal alcohols is very useful in the synthesis of chemicals and fuels.

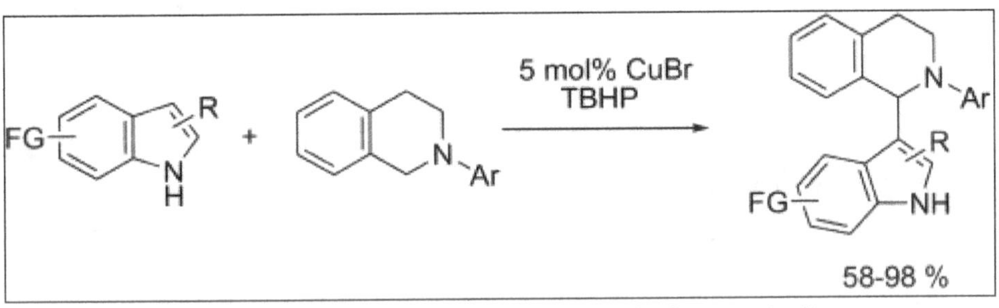

CDC reactions of various indoles with tetrahydroisoquinolines.

However, the direct conversion of C–H bonds into C–C bonds leads to more efficient syntheses of complex products with reduced synthetic operations. Recently, great progress has been made in transition-metal-catalyzed activation and further reaction of C–H bonds. Li and others have developed various methods to generate C–C bonds directly from two different C–H bonds in the presence of an oxidizing reagent through a cross-dehydrogenative coupling (CDC) catalyzed by transition metals. For example, (NH)-indoles and tetrahydroisoquinolines were converted directly into alkaloids by using such a coupling. Recently, an elegant cross-coupling of two aryl C–H bonds to form arene–arene coupling products was reported by Fagnou, Sanford, and others.

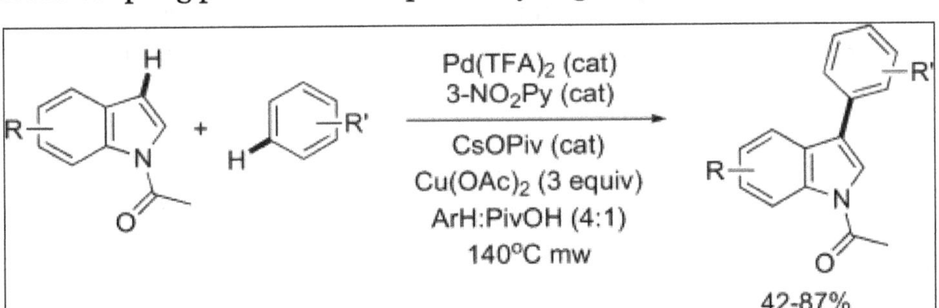

Synthesis without Protections

Because of the nature of classical chemical reactivity, organic synthesis extensively utilizes protection–deprotection of functional groups, which increases the number of steps in synthesizing the desired target compounds. Novel chemistry is needed to perform organic synthesis without protection and deprotection. Recently, progress has been made on this subject. For example, Baran *et al.* have reported a total synthesis of a natural product without any protecting groups. Another instance is the efficient synthesis of (+)-3-deoxy-d-glycero-d-galacto-nonulosonic acid (KDN) by using the indium-mediated allylation reaction in water reported by Chan and Li.

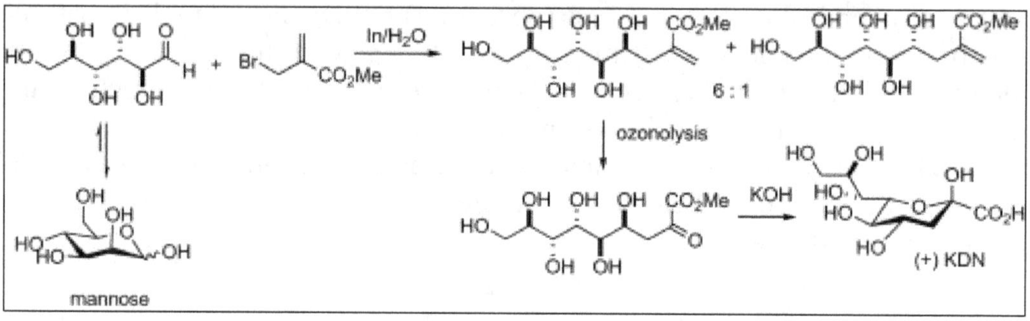

The Knoevenagel condensation of the β-diketone with hemiacetalic sugar gave β-*C*-glucosidic ketone in water directly.

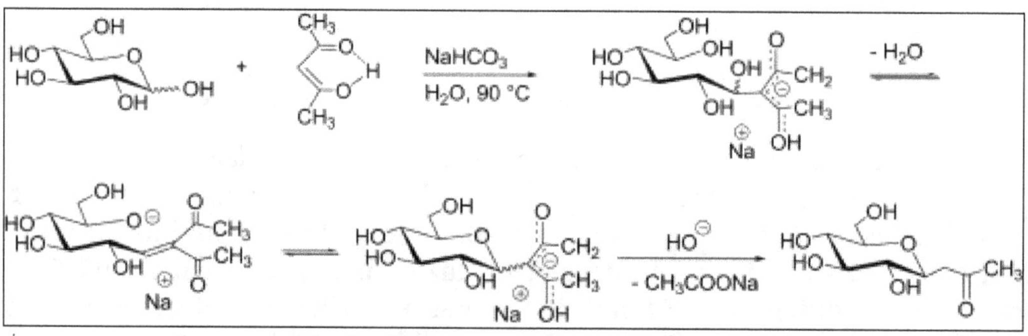

One step synthesis of β-C-glucosidic ketone in aqueous medium.

The Click chemistry developed by Sharpless tolerates a wide range of functionalities and allows the direct modification of biological compounds. The archetypical example of Click chemistry is the Huisgen 1,3-dipolar cycloaddition of alkynes to azides to form 1,4-disubsituted 1,2,3-triazoles catalyzed by Cu(I). The reaction is mild and highly efficient, and does not require protecting groups.

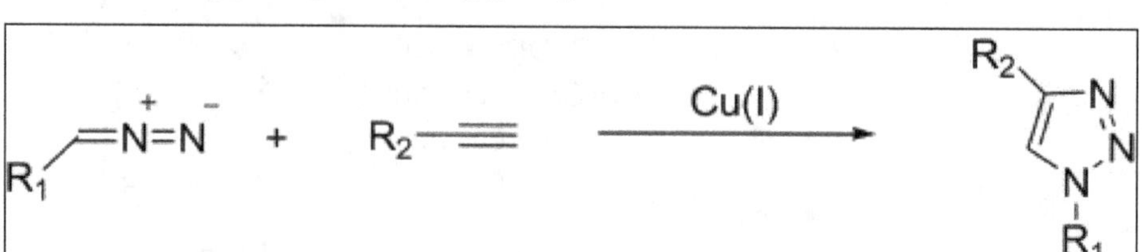

Tandem/Cascade/Flow Reactors

Also of fundamental importance to greener syntheses is the development of tandem and cascade reaction processes that incorporate as many reactions as possible to give the final product in one operation. For example, a palladium-catalyzed tandem reaction resulted in multirings in one step. Another example is Jamison's synthesis of the core piece of "ladder" polyether marine natural products through a biomimetic cascade cyclization in neutral water.

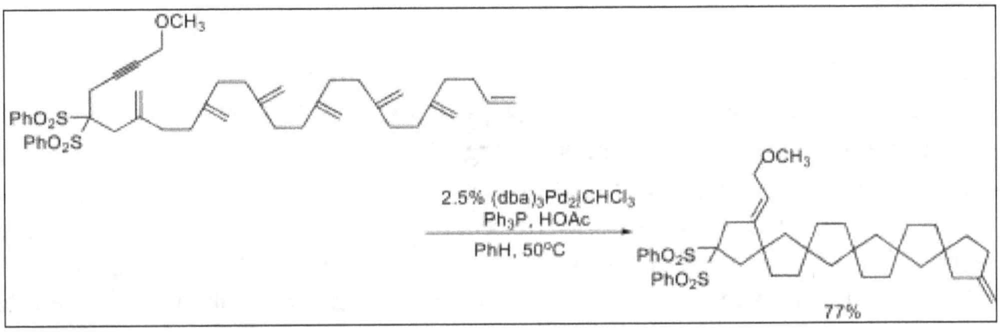

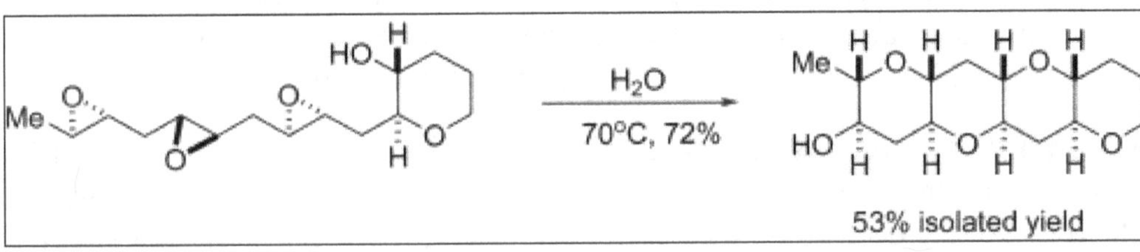

An alternative way to simplify organic syntheses into a single operation is to perform sequential reactions in a flow reactor. Ley and colleaques reported that a multistep synthesis of the alkaloid natural product (±)-oxomaritidine can be accomplished by using microfluidic pumping systems that pass material through various packed columns containing immobilized reagents, catalysts, scavengers, or catch-and-release agents, combining seven separate synthetic steps linked into one continuous sequence.

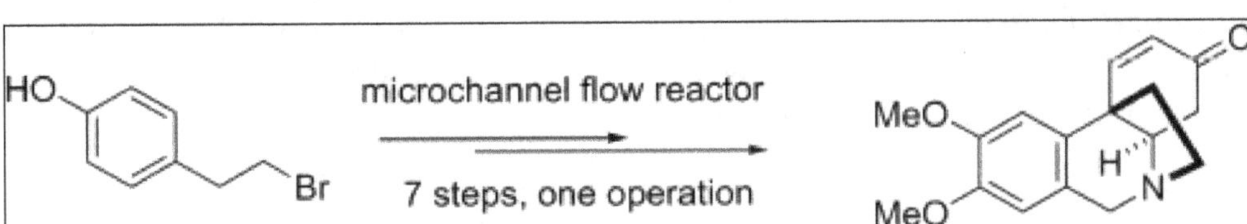

Synthesis in flow of (±)-oxomaritidine.

Biocatalysis

Through millions of years of evolution and "sustainability," nature developed highly efficient and selective means to achieve the desired transformations. The potential usefulness of various catalysts of Nature, such as enzymes, whole cells, and catalytic antibodies for organic synthesis, has become more and more recognized. Frequently, biocatalysis leads to extremely high reaction rates and selectivities such as enantioselectivities that go beyond the reach of chemical catalysts. These developments have provided powerful and parallel tools in the synthetic chemist's toolbox. However, the high substrate specificity of enzymes presents a dilemma for synthetic chemistry in which wide substrate applicability is desired. The recent exciting development in "directed evolution" provides potential opportunities in using biological catalysts to overcome this issue.

Solvents

Solvents are auxiliary materials used in chemical synthesis. They are not an integral part of the compounds undergoing reaction, yet they play an important role in chemical production and synthesis. By far, the largest amount of "auxiliary waste" in most chemical productions is associated with solvent usage. In a classical chemical process, solvents are used extensively for dissolving reactants, extracting and washing products, separating mixtures, cleaning reaction apparati, and dispersing products for practical applications. Although the invention of various exotic organic solvents has resulted in some remarkable advances in chemistry, the legacy of such solvents has led to various environmental and health concerns. Consequently, as part of Green Chemistry efforts, various cleaner solvents have been evaluated as replacements.

The primary function of solvents in classical chemical syntheses is to facilitate mass transfer to modulate chemical reactions in terms of reaction rate, yields, conversions, and selectivity. They do this by dissolving the reactants in dilute homogeneous mixtures. The ironic aspect of this process is that, after the reaction, the final product has to be reseparated from the solvent through energy-intensive means.

The development of Green Chemistry redefines the role of a solvent: An ideal solvent facilitates the mass transfer but does not dissolve. In addition, a desirable green solvent should be natural, nontoxic, cheap, and readily available. More desirably, it should have additional benefits of aiding the reaction, separation, or catalyst recycling. The concept of organic reactions "on-water," in which non-water-soluble reagents react by floating on water to generate a non-water-soluble and readily separable final product, is an excellent manifestation of this new definition.

Water

The only natural solvent on earth is water. Life requires the construction of chemical bonds in an aqueous environment. It is obvious that water is the most inexpensive and environmentally benign solvent. Since it was reported that Diels–Alder reactions could be greatly accelerated by using water as a solvent instead of organic solvents, there has been considerable attention dedicated to the development of organic reactions in water. Besides Diels–Alder reactions, other examples cover almost all of the most useful organic reactions, even "water-sensitive" reactions. In many cases, because of hydrophobic effects, using water as a solvent not only accelerates reaction rates but also enhances reaction selectivities, even when the reactants are sparingly soluble or insoluble in the medium. Furthermore, the low solubility of oxygen gas in water, an important property in the early development of life in an anaerobic environment, can facilitate air-sensitive transition-metal catalysis in open air. The use of water as a solvent also implies the elimination of tedious protection–deprotection processes for certain acidic-hydrogen-containing functional groups, which contributes to the overall synthetic efficiency. Water-soluble compounds, such as carbohydrates, can be used directly without the need for laborious derivatization and water-soluble catalysts can be reused after separation from water-insoluble organic products. Aqueous organic chemistry is also essential in the emerging field of chemical biology, which uses chemical tools to study biological systems.

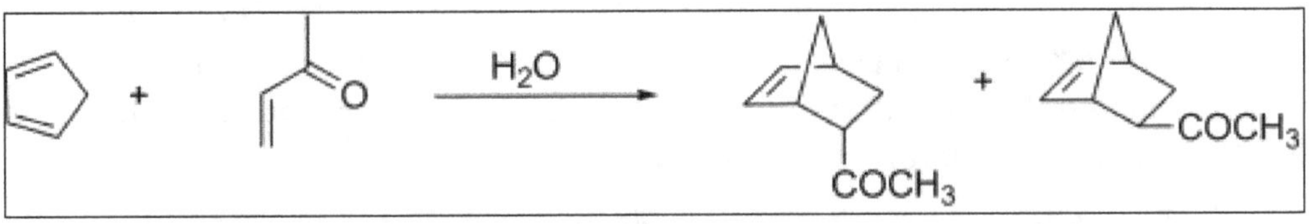

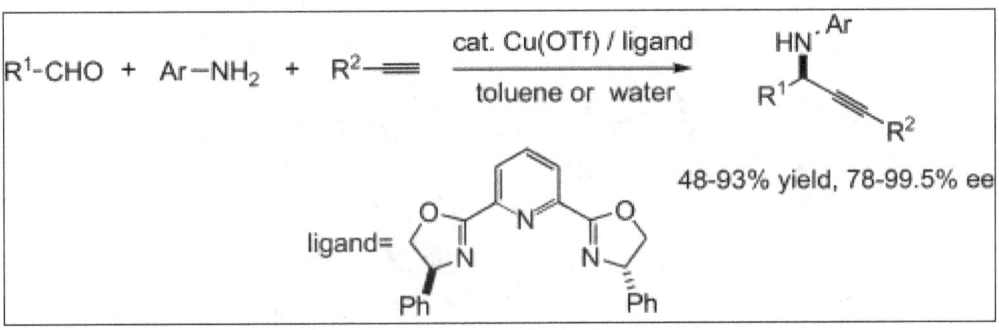

One challenge of using water as a solvent is the regeneration of pure water that contains only minor impurities. In this respect, newer purification technologies such as ultrafiltration or natural evaporation (if the impurity is not vaporizable) help. Another challenge of using water is separating water-soluble products from water. In addition, many organic compounds are not soluble in water. Although "on-water" techniques have provided excellent solutions for some situations, there will be cases where completely soluble in water is desired.

CO_2

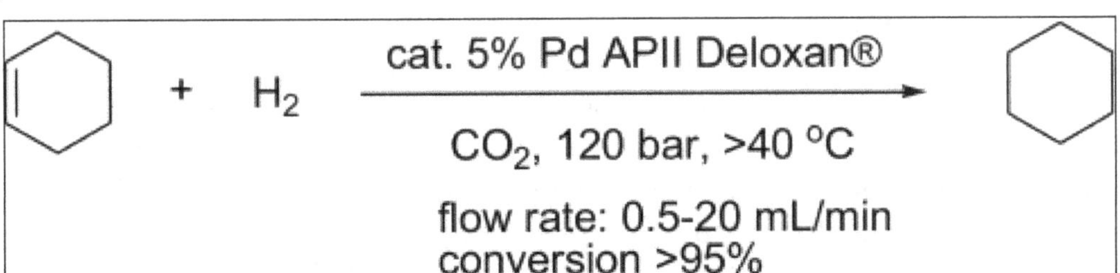

In some cases, water is undesirable. Although some chemical processes may be modified to use water, green solvents with different properties than water are nevertheless needed. One such solvent is liquid and supercritical CO_2. It is also a natural solvent, although some energy (pressure) is consumed in its production. In addition, CO_2 is renewable, nonflammable, and readily evaporating. Other excellent features of CO_2 include its fast drying time, better ability to dissolve organic compounds, and better flow ability because of its low viscosity compared with other solvents including water. These properties are complementary to water and provide supplementary needs. Through the landmark work by Noyori et al., DeSimone et al., Tumas and colleaques, Leitner, and others, supercritical CO_2 is emerging as an important medium for chemical synthesis. One special feature of liquid and supercritical CO_2 is its high mixibility with gases, which offers high efficiency (and often higher selectivity) in reactions such as hydrogenations with hydrogen gas and oxidations with air. Another feature of CO_2 is its rapid separation from catalysts and products by simple depressurization and recapture. Taking the advantage of these features, Britz *et al.* and Cole-Hamilton *et al.* developed an efficient flow chemical reactor for clean synthesis.

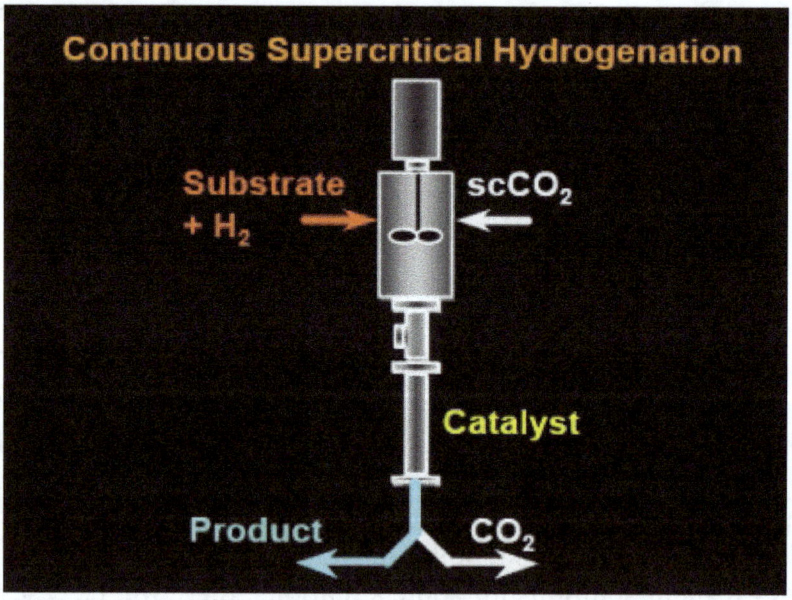

Continuous supercritical hydrogenation.

Nonnatural Solvents

In addition to the two "natural green solvents", various nonnatural ones have also been intensively studied as green alternatives. The most widely studied ones are ionic liquids. The greatest advantage of these solvents is their low vapor pressures, which offer advantages in reducing volatile organics in the air. Such novel solvents also offer various interesting new chemistries such as dissolving cellulose and changing the outcome of reactions. Another innovative discovery is the recently developed "switchable solvents" by Jessop, Liotta, Erckert, and others. Such solvents change their properties with different needs. Beside these solvents, other synthetic solvents such as fluorous and property-changing soluble polymer systems have been evaluated as potential green alternatives.

Green Chemistry Metrics

Green chemistry metrics are metrics that measure aspects of a chemical process relating to the principles of green chemistry. These metrics serve to quantify the efficiency or environmental performance of chemical processes, and allow changes in performance to be measured. The motivation for using metrics is the expectation that quantifying technical and environmental improvements can make the benefits of new technologies more tangible, perceptible, or understandable. This, in turn, is likely to aid the communication of research and potentially facilitate the wider adoption of green chemistry technologies in industry.

For a non-chemist the most attractive method of quoting the improvement might be *a decrease of* X unit cost per kilogram of compound Y. This, however, would be an over-simplification—for example, it would not allow a chemist to visualise the improvement made or to understand changes in material toxicity and process hazards. For yield improvements and selectivity increases, simple percentages are suitable, but this simplistic approach may not always be appropriate. For example, when a highly pyrophoric reagent is replaced by a benign one, a numerical value is difficult to assign but the improvement is obvious, if all other factors are similar.

Numerous metrics have been formulated over time and their suitability discussed at great length. A general problem observed is that the more accurate and universally applicable the metric devised, the more complex and unemployable it becomes. A good metric must be clearly defined, simple, measurable, objective rather than subjective and must ultimately drive the desired behavior.

Effective Mass Yield

Effective mass yield is defined as the percentage of the mass of the desired product relative to the mass of all non-benign materials used in its synthesis. Hudlicky *et al.* suggests the following equation:

Effective mass yield (%) = mass of products × 100%/mass of non-benign reagents.

This metric requires further definition of a benign substance. Hudlicky defines it as "those by-products, reagents or solvents that have no environmental risk associated with them, for example, water, low-concentration saline, dilute ethanol, autoclaved cell mass, etc.". This definition leaves the metric open to criticism, as nothing is non-benign (which is a subjective term) and the substances listed in the definition have some environmental impact associated with them. The formula also fails to address the level of toxicity associated with a process. Until all toxicology data is available for all chemicals and a term dealing with these levels of "non-benign" reagents is written into the formula the effective mass yield is not the best metric for chemistry.

Carbon Efficiency

Carbon efficiency is how much carbon ends up in the useful product compared to how much carbon was used to create the product.

> Carbon efficiency (%) = amount of carbon in product × 100%/total carbon present in reactants.

This metric is a good simplification for use in the pharmaceutical industry as it takes into account the stoichiometry of reactants and products. Furthermore, this metric is of interest to the pharmaceutical industry where development of carbon skeletons is key to their work.

Reaction Mass Efficiency

The reaction mass efficiency takes into account atom economy, chemical yield and stoichiometry. For a generic reaction A + B → C, the formula can take either of the two forms:

- Reaction mass efficiency = molecular weight of product C × yield/m.w. A + (m.w. B × molar ratio B/A).
- Reaction mass efficiency = mass of product C × 100%/mass of A + mass of B.

The conservation of mass principle dictates that the total molecular mass of the reactants is the same as the total molecular mass of the products. In an ideal chemical process, the amount of starting materials or reactants equals the amount of all products generated and no atom is wasted. However, in some processes, some of the consumed reactant atoms do not become part of the intended products. This can be a concern when raw materials are expensive or when economic and environmental costs of disposal of the waste are high. Like carbon efficiency, this measure shows the "clean-ness" of a reaction but not of a process. Neither metric takes into account waste produced. For example, these metrics could present a rearrangement as "very green" but fail to address any solvent, work-up, and energy issues that make the process less attractive.

Environmental (E) Factor

The first general metric for green chemistry remains one of the most flexible and popular ones. Roger A. Sheldon's E-factor can be made as complex and thorough or as simple as desired and useful.

The E-factor the ratio of the mass of waste per mass of product:

> E-factor = total waste/product

As examples, Sheldon calculated E-factors of various industries:

Table: E-Factors across the chemical industry.

Industry sector	Annual production (t)	E-factor	Waste produced (t)
Oil refining	10^6-10^8	Ca. 0.1	10^5-10^7
Bulk chemicals	10^4-10^6	<1–5	10^4-5×10^6
Fine chemicals	10^2–10^4	5–50	5×10^2–5×10^5
Pharmaceuticals	10–10^3	25–100	2.5×10^2–10^5

It highlights the waste produced in the process as opposed to the reaction, thus helping those who try to fulfil one of the twelve principles of green chemistry to avoid waste production. E-factors ignore recyclable factors such as recycled solvents and re-used catalysts, which obviously increases the accuracy but ignores the energy involved in the recovery (these are often included theoretically by assuming 90% solvent recovery). The main difficulty with E-factors is the need to define system boundaries, for example, which stages of the production or product life-cycle to consider before calculations can be made.

Crucially, this metric is simple to apply industrially, as a production facility can measure how much material enters the site and how much leaves as product and waste, thereby directly giving an accurate global E-factor for the site. Table shows that oil companies produce a lot less waste than pharmaceuticals as a percentage of material processed. This reflects the fact that the profit margins in the oil industry require them to minimise waste and find uses for products which would normally be discarded as waste. By contrast the pharmaceutical sector is more focused on molecule manufacture and quality. The (currently) high profit margins within the sector mean that there is less concern about the comparatively large amounts of waste that are produced (especially considering the volumes used) although it has to be noted that, despite the percentage waste and E-factor being high, the pharmaceutical section produces much lower tonnage of waste than any other sector. This table encouraged a number of large pharmaceutical companies to commence "green" chemistry programs.

By incorporating yield, stoichiometry and solvent usage the E-factor is an excellent metric. Crucially, E-factors can be combined to assess multi-step reactions step by step or in one calculation.

The EcoScale

The EcoScale metric was proposed for evaluation of the effectiveness of a synthetic reaction. It is characterized by simplicity and general applicability. Like the yield-based scale, the EcoScale gives a score from 0 to 100, but also takes into account cost, safety, technical set-up, energy and purification aspects. It is obtained by assigning a value of 100 to an ideal reaction defined as "Compound A (substrate) undergoes a reaction with (or in the presence of)inexpensive compound(s) B to give the desired compound C in 100% yield at room temperature with a minimal risk for the operator and a minimal impact on the environment", and then subtracting penalty points for non-ideal conditions. These penalty points take into account both the advantages and disadvantages of specific reagents, set-ups and technologies.

BioLogicTool plots

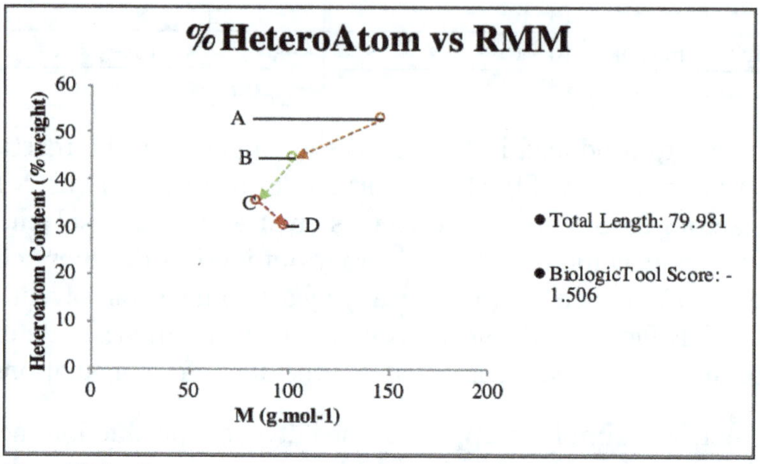

Plots of %heteroatoms against molar mass so called BiologicTool plots.

This freely available tool offers a visual representation of a chemical route from data added by the user (starting material, intermediates and product names, their chemical formula, molar masses and optionally yields of reaction steps). The plots as well as the two scores given were designed to help assessing the rationality of a chemical route with special emphasis on the comparison of bio-based routes with petrol-based ones.

The percentage by mass of heteroatoms contained in the starting material, intermediates, and final product is plotted against their respective molar masses. The visual representation of the chemical routes studied comes with two scores namely, the Total Length and the BioLogictool Score. After normalization of the data, the Total Length is calculated by summation of the individual length of all plotted vectors (commencing from starting material and finishing to intermediates/products). The BioLogicTool score is then obtained after the division of the Total Length by the length of a hypothetical direct vector starting from the original feedstock (A in the example) and finishing to the product (D). The closer to 1 the Biologictool score is and the smaller the Total Length, the more rational the chemical route will be.

Green Chemistry's 12 Principles

Twelve principles of green chemistry have been developed by Paul Anastas and John Warner of EPA, and in their Green Chemistry Theory and Practice book, 1998, they explained their meaning in practice. The principles of green chemistry speak about the reduction or removal of dangerous or harmful substances from the synthesis, production and application of chemical products and thus the use of substances dangerous to human health and the environment is reduced or eliminated. When designing a green chemistry process, it is impossible to meet the requirements of all twelve principles of the process at the same time, but it attempts to apply as many principles as possible during certain stages of synthesis.

Prevention

It is better to prevent waste than to treat or clean up waste after it has been created. The first principle of green chemistry is the principle of prevention, i.e. the prevention of waste generation, which is more favorable for humans and the environment, and ultimately cheaper than treating waste and destroying it after it has emerged.

The justification for the introduction of this principle is confirmed by the fact that around 12 billion tonnes of waste, or about 300 million tonnes of hazardous waste for human health and the environment (so-called hazardous waste) are produced annually in the United States. The chemical industry produces 70 percent of the total amount of hazardous waste and the highest organic toxic waste (about 150,000 tons) with methanol and xylenes prevailing. Waste sanitation “consumes about 2.2 percent of gross US product, and costs continue to grow”. Organic waste, which is harmful to humans and the environment, is primarily produced at certain stages of synthesis, so-called. “Dirty reactions” during which toxic reactants and solvents are used, and due to the harsh reaction conditions, a large number of toxic byproducts are formed. These are the most common basic reactions of organic synthesis (halogenation, oxidation, alkylation, nitration and sulfonation) that are applied in different industrial branches.

Although the chemical industry, as well as other chemical manufacturers, have long ago avoided prevention, the interest of green chemistry and the community is precisely to prevent waste generation. However, the absolute prevention of waste generation in practice is virtually impossible since no input raw material can be fully utilized. On the other hand, one waste disposed of represents the final loss of material goods in the circular flow: Production-consumption. Therefore, any return of material goods to a circular stream represents a pure economic gain and it is necessary to think first whether it is possible to prevent the generation of waste and if it is not necessary to devise the way in which the amount of waste produced in production can be utilized in the best possible way, so it becomes useful.

This approach to the problem has brought positive results: the paint and varnish industry already produces solvent-free paints and lacquers. The detractors industry has already thrown out all phosphorus-containing detergents. Asbestos is no longer used in practice.

Along with finding new medicines, one of the biggest challenges of the pharmaceutical industry is to design and apply procedures that have minimal environmental impact in the manufacturing process. This primarily means reducing the amount of waste generated in production processes. Since nearly 80% of waste in the pharmaceutical industry is associated with solvents, and that about 60% of their energy is consumed, the solution is to reduce the use of solvents. A good example is the new "green" production process of sertraline by which the introduction of ethanol as the sole solvent eliminates the need for the use, distillation and recovery of four solvents (methylene chloride, tetrahydrofuran, toluene and hexanes) resulting in a reduction in solvent consumption of 250 to 25 liters per kg of sertraline.

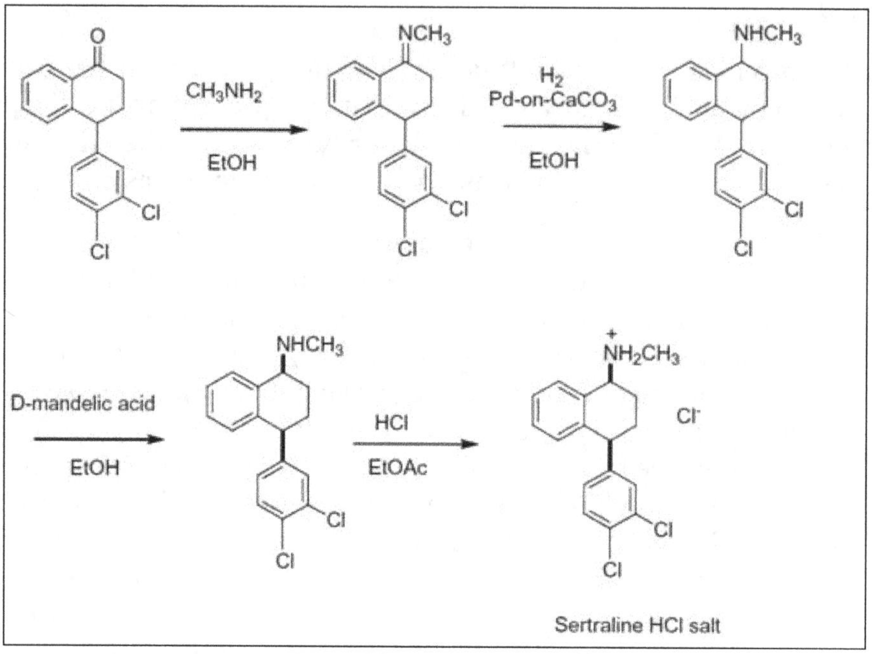

"Green" hesis of sertraline.

Atom Economy

Synthetic methods should be designed to maximize the incorporation of all materials used in the process into the final product. The principle of Atom Economy is logically linked to the principle of waste prevention, since it requires all raw materials used in production to maximize utilization or inclusion in the final product to ultimately reduce the amount of waste. This means that the chemical synthesis should be designed in such a way that the final product maximizes the input of raw materials or design such synthetic products that will use the entire material used for synthesis in the final product.

The principle of increasing atomic usability was defined in 1991 by Barry Trost of Stanford University. Trost believes that introducing the concept of usability atoms is essentially the prevention of waste at the molecular level. Barry Trost's concept initiated the redesign of existing synthetic reactions until then established on the principle of "making a product regardless of price". These modifications are useful and because they generally lead to increased yields.

There is a known progress in the synthesis of ibuprofen. The main problem of old synthesis (boots process) is low economic cost, because the utilization of input raw materials is only about 40%. In the 1990s a new "green" method of ibuprofen synthesis was developed, involving only three steps, and almost all transitional materials were converted to the product (up to 99%) or regenerated and returned in the process and almost almost eliminated the generation of waste materials And this process is one of the processes of "green synthesis".

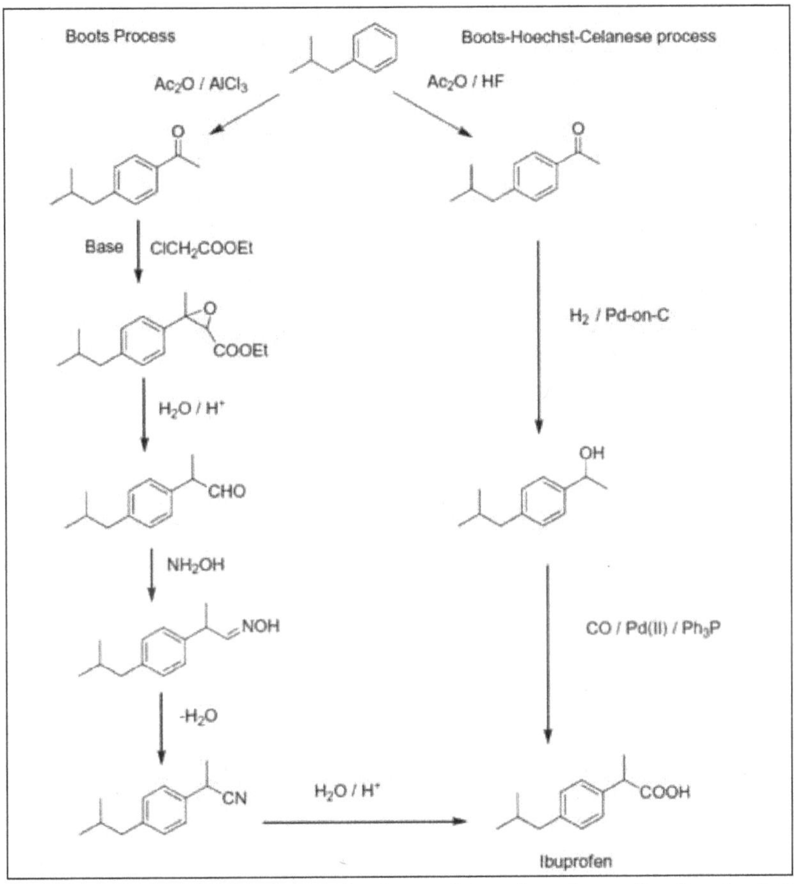

Comparison of the Boots Process and the "Green" Hoechst Ibuprofen Synthesis Process.

Figure shows a comparative comparison of classic Boots and the "green" Hoechst synthesis of ibuprofen. The new process has much higher atomic efficiency and almost no waste (waste materials are recycled in the process) thus contributing to pollution prevention. Atom economy is defined as the ratio of relative molecular masses of the desired product and all reactants expressed in percentages.

% of atomic efficiency = (Mr of the desired product/Mr of all reactants) x 100

The Boots synthesis consists of six synthetic steps and yields 40%, while the Hoechst process of obtaining ibuprofen consists of only three steps and yields a reaction of 77%, namely 99% if the fact that the acetic acid is formed in the first. The synthesis step is extracted and reused in production. That is why Hoechst is a more economical and greener way to get ibuprofen. In Tables, a comparison of the Boots process and the "green" Hoechst process of ibuprofen synthesis according to atomic viability.

Tables: Effectiveness of Atoms in the Green "Hoechst" Process of Ibuprofen Synthesis.

Reactant		Used in Ibuprofen		Not used in ibuprofen	
Formula	Relative molecular mass (M_r)	Formula	Relative molecular mass (M_r)	Formula	Relative molecular mass (M_r)
$C_{10}H_{14}$	134	$C_{10}H_{13}$	133	H	1
$C_4H_6O_3$	102	C_2H_3	27	$C_2H_3O_3$	75
$C_4H_7ClO_2$	122,5	CH	13	$C_3H_6ClO_2$	109,5
C_2H_5ONa	68		0	C_2H_5ONa	68
H_3O	19		0	H_3O	19
NH_3O	33		0	NH_3O	33
H_4O_2	36	HO_2	33	H_3	3
Total		Ibuprofen		Production waste	
$C_{20}H_{42}NO_{10}ClNa$	514,5	$C_{13}H_{18}O_2$	206	$C_7H_{24}NO_8ClNa$	308,5

% atom economy = (206/514,5) x 100 =40%

Reactant		Used in Ibuprofen		Not used in ibuprofen	
Formula	Relative molecular mass (M_r)	Formula	Relative molecular mass (M_r)	Formula Relative molecular mass (M_r)	
$C_{10}H_{14}$	134	$C_{10}H_{13}$	133	H	1
$C_4H_6O_3$	102	C_2H_3	43	$C_2H_3O_2$	59
H_2	2	H_2	2		0
CO	28	CO	0		0
Total		Ibuprofen		Production waste	0
$C_{15}H_{22}NO_4$	266	$C_{13}H_{18}O_2$	206	$C_2H_4O_2$	60

% atom economy = (206/266) x 100 =77%

The E-factor is used to compare the process of comparing the proportions of waste materials and the desired product. The calculation of the E-factor is determined by the ratio of mass of waste (kg) per unit of product in kilograms.

E-factor = kg of waste / kg of product

And represents "the actual amount of waste generated in the process, including the losses of solvents, acids and bases used in the work and, in principle, waste from energy production". Waste includes products that do not have any further use, but also reagents and solvents used during production and are not recycled or recycled.

Sheldon points out that the use of E-factors varies greatly between the chemical industry and states that the "good" E factor usually spends around 0.1, meaning that 10 kg of the desired product is only 1 kg of waste and by-products. In pharmaceutical production, when essential for high purity products, the E-factor can be 100, which means that every kg of product produces 100 kg of waste. However, the actual amount of waste also depends on the quantities of the product. Even with a much lower E-factor, the oil industry produces higher amounts of waste than pharmaceutical production where the E-factor is high but is significantly lower in production. Table shows E-factors of various sectors of the chemical industry.

Table: E-factor in chemical industry.

Industry	Product (t)	E-factor	Waste (t)
Rafinery	10^6 - 10^8	cca 0,1	10^5 - 10^7
Salts	10^4 - 10^6	< 1- 5	10^4 - 5×10^6
Fine chemicals	10^2 - 10^4	5 - 50	5×10^2 - 5×10^5
Pharmaceuticals	10 - 10^3	25 - 100	2.5×10^2 - 10^5

Less Hazardous Chemical Syntheses

Wherever practicable, synthetic methods should be designed to use and generate substances that possess little or no toxicity to human health and the environment. Most chemical synthesis reactions, which usually take place in multiple stages, use toxic reagents. Although the product does not contain these toxic substances, there is a risk of its contamination and redesigning these processes is a task of green chemistry.

The Less Hazardous Chemical Synthesis advocates, wherever possible, the creation of synthetic methods for the use and creation of substances that are little or no toxic to human health and the environment. Replacing harmful chemicals with biological enzymes makes many industrial processes cleaner and cheaper.

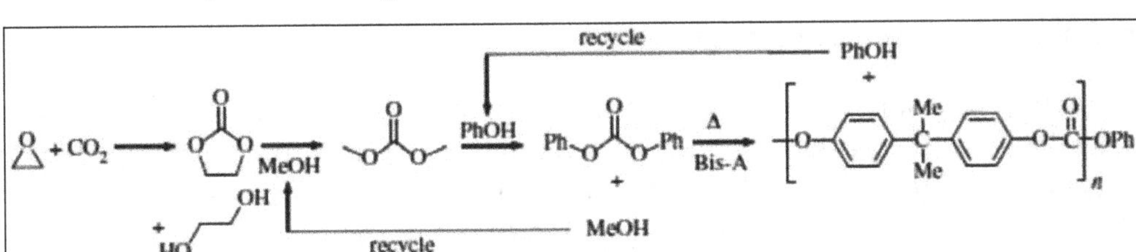

Synthesis of polycarbamate.

As an example, a new Asahi Kasei's polycarbamate synthesis (PC) process is conceptually simple, based on the substitution of toxic carbonyl dichloride (COC_2) with CO_2. This process also results in the removal of dichloromethane (CH_2Cl_2) as a solvent. The total reaction consists of ethylene oxide (CH_2) 20, CO_2 and bisphenol-A $(C_{15}H_{16}O_2)$ to give polycarbamate and ethylene glycol $C_2H_6O_2$.

Designing Safer Chemicals

Chemical products should be designed to affect their desired function while minimizing their toxicity. Minimizing toxicity, while maintaining function and efficiency, can be one of the most challenging aspects of designing safer products and processes, and achieving that goal requires understanding not only chemistry but also the principles of toxicology and environmental science.

Designing Safer Chemicals advocates the design of chemical products in a way that reduces their toxicity and maintains their effectiveness. The goal of producing safe chemicals (non-carcinogenic, mutagenic, neurotoxic) is the balance between optimal performance and chemical product function, ensuring that toxicity and risk are reduced to the lowest possible level. In other words, the use of toxic chemicals should be avoided and replaced inhospitable wherever possible, but should take account of their efficacy.

This principle is used in the development of new insecticides and pesticides that are specific to target organisms, i.e. they are toxic only to target organisms and decompose into environmentally harmless substances. Another example is the use of highly toxic organic tin-based organic compounds (Sn), previously coated on the outside to prevent the capture of seaweed and plankton. These organic compounds have been replaced by a product called Sea-Nine which is completely degradable and non-toxic. Also, an example is the development of oxidant activator for hydrogen peroxide. This allows replacement of chlorine bleaches (preparations that damage the ozone layer) by hydrogen peroxide in paper production. An example is the production of polymers of polyphenylene sulfone (PPSU), which is now widely used for indoor airplanes and is also introduced in underground trains where it is also important to use non-flammable materials. It is a new engineering plastic characterized by a unique combination of useful environmental, mechanical, and flame resistant properties.

Safer Solvents and Auxiliaries

The use of auxiliary substances (e.g., solvents, separation agents, etc.) should be made unnecessary wherever possible and innocuous when used. Chromatographic separations, where large quantities of solvents are used, are problematic due to environmental pollution. Most conventional organic solvents are toxic, flammable and corrosive. Their recycling is linked to energy efficient distillation with considerable losses and therefore the development of environmentally-friendly solvents is necessary.

Safer Solvents and Auxiliaries recommends that the process of synthesis be maximally reduced and, whenever possible, avoid the use of auxiliary chemical substances (e.g. solvents, separating agents, etc.) when used they should be harmless. According to the principles of green chemistry, the choice of suitable substitutions for organic solvents is based on: worker safety, process safety, environmental safety and sustainability of the process. The solvent should be chemically and physically stable, low volatility, easy to use and easy to recycle. According to the suitability of use, conventional solvents are suitable, usable, and undesirable.

Table: Solvent selection according to usability.

Suitable	Usable	Undesirable
Methanol	Cyclohexane	Pentane
Ethanol	Methylcyclohexane	Hexane
Propan-1-ol	Heptane	Diisopropyl ether
Propan-2-ol	Isooctane	Dichloromethane
Butan-1-ol	T-Butyl methyl ether	Chloroform
T-butanol	Acetonitrile	Benzene
Ethyl acetate	Tetrahydrofuran	Dimethyl acetate
Isopropyl acetate	Acetic acid	Carbon tetrachloride
Acetone	Xylene	Dimethoxyethane
Methyl ethyl ketone	Toluene	Pyridine

For now, it is promising to replace the known organic solvents with recyclable solvents, which are ionic liquids - salts at room temperature in the liquid state. Unlike volatile organic compounds, ionic liquids have low vapor pressure, do not vaporise and do so easily, resulting in safer chemical processes.

Design for Energy Efficiency

Energy requirements of chemical processes should be recognized for their environmental and economic impacts and should be minimized. If possible, synthetic methods should be conducted at ambient temperature and pressure.

The oil crisis in 1973 has initiated the development of a number of processes in which energy savings are taken into account, with the aim of exploiting every kJ of energy in the production process. Following the above-mentioned Principle of Energy Efficiency, whose other name is Design for Energy Efficiency, as a fundamental requirement, minimizes the use of energy. Table shows possible ways to improve energy efficiency in the chemical industry.

Table: Improving Energy Efficiency in the Chemical Industry.

Maintenance and recovery	Good insulation and well-maintained equipment reduce heat and energy losses. Waste generated by chemical processes often has an energy value and can be converted into useful fuel for office heating and hot water production. In some cases this heat can be shared with the local community.

Chemical reactions - choice and conditions	It is necessary to choose the reactions and catalysts that require lower temperatures. In order to minimize energy requirements, the principle of increasing energy efficiency advocates the process of synthesis at room temperature and atmospheric pressure, whereby the use of catalyst is of great use. Catalysts are developed so that the process can be run at lower temperatures and pressures (high temperatures and pressures require enough energy).
Combined Heat and Power (CHP)	Production plants often generate their own cogeneration (Combined Heat and Power or CHP), which involves simultaneous production of electrical and useful heat energy in a single process. It is more efficient because it eliminates the loss in transmission, And the excess heat released during the cogeneration process can be used on site for various purposes.

Tomatoes that grow in a greenhouse that is heated by the use of waste steam from a nearby chemical plant for the production of ammonia. Since CO_2 concentrations in greenhouses fall below 50%, plant growth can be achieved, and by extrusion from greenhouse gas, carbon dioxide is used as a tomato seed growth bi-activator.

Use of Renewable Feedstocks

A raw material or feedstock should be renewable rather than depleting whenever technically and economically practicable. The seventh principle of green chemistry advocates Use of Renewable Feedstocks wherever it is technically and economically acceptable. For example, it is more convenient to use renewable raw materials than a variety of plastic materials, and then to waste away the waste materials. Because of this, the making of biodegradable plastic materials is a current trend. Biodegradable packaging has a future in the food industry. Numerous factors, including politics and changes in legislation, as well as global demand for food and energy resources, certainly affect the development of biodegradable packaging. The principle also implies the use of renewable energy technologies that include solar energy, wind power, hydropower, biomass energy and biofuels. Over the past ten years, significant progress has been made in the development of fuels, chemicals and materials from renewable raw materials. For example, Brazil with its sugar cane production and bioethanol production ensures energy independence and employment while Europe's biodiesel is produced from rapeseed oil.

In the case of bioplastics, the use of renewable raw materials in production positively affects energy consumption and CO_2 emissions. Coca-Cola, a world-widescale company for the time being, manufactures bottles made of 30% polyethylene (PE) blends, while American company NatureWorks uses bottles made from lactic acid polymers (PLA) made from lactic acid, obtained by fermentation of dextrose obtained from starch, Most commonly corn. About 1 kg of PLA requires about 2.5 kg of corn.

Reduce Derivatives

Unnecessary derivatization (use of blocking groups, protection/deprotection, temporary modification of physical/chemical processes) should be minimized or avoided if possible, because such steps require additional reagents and can generate waste. One of the key principles of green chemistry in the synthesis of target molecules is to avoid the use of chemical derivatives (Reduce Derivatives). The principle advocates, wherever possible, the avoidance of physical-chemical processes in which blocking and unblocking of appropriate groups during the synthesis are used, that is, whenever possible, the biological processes in the synthesis should be used to avoid synthesis of the products for which they are not. There are enzymes to degrade them. If possible, it is necessary to reduce or avoid unnecessary derivatization (group blocking, protection/elimination protection, temporary physicochemical modification) because such steps require additional reagents and can generate waste.

Typical example is the production of antibiotics based on penicillin or replacement of classical chemical enzymatic processes whereby the 6-aminopenicillic acid is obtained by reacting with the catalyzed immobilized enzyme penicillin amide. This resulted in several chemical steps being replaced by an enzymatic reaction, and no longer required a low temperature (-60 °C), organic solvents, and completely unsuitable conditions that increased and complicated production in the case of chemical synthesis.

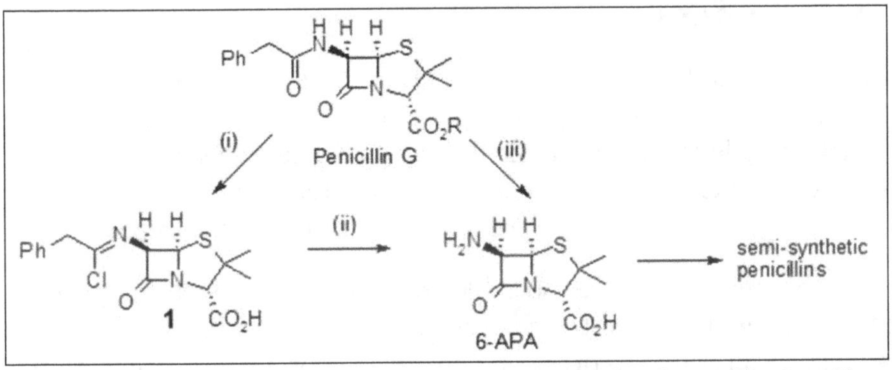

Synthesis of 6-aminopenicilic acid catalyzed by immobilized penicillin G amide.

Catalysis

Catalytic reagents (as selective as possible) are superior to stoichiometric reagents. In order to protect the environment, the catalysis principle promotes the use of biodegradable catalysts, which imply less energy use, avoiding the use of organochlorine compounds and reducing the use of water or less waste water.

Like all catalysts, enzymes function in a way that lowers the activation energy of an individual reaction, and thus accelerates, up to several million times. In doing so, the enzyme remains unchanged throughout the duration of the reaction to which it affects, and this allows it to become completely unchanged when the reaction comes to an end. Also, enzymes do not affect the relative energy between the reactants and the products, nor to the related reactions. However, what enzymes stand out among all other catalysts is their specificity in terms of stereochemistry, chemical selectivity and specificity. Compared with non-biological catalysts, biocatalysts have a great advantage given the rate of reaction, catalytic specificity, lower cost, etc., but lack of heat sensitivity and poor stability.

The classic catechol synthesis is derived from benzene (non-renewable feedstocks) in several reaction stages requiring severe reactions under which undesirable byproducts are produced. It is replaced by a biocatalytic synthesis of D-glucose (renewable raw materials) in the presence of genetically modified Escherichia coli and is performed in one reaction stage, no byproducts and production is economically viable.

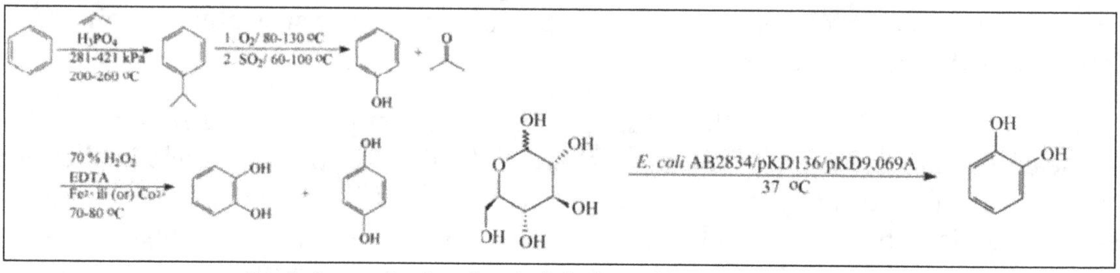

Catehole synthesis - classical (left) and biocatalytic (right).

Design for Degradation

Chemical products should be designed so that at the end of their function they break down into innocuous degradation products and do not persist in the environment. The principle of creating degradable chemicals and products or design for degradation demands the creation of chemical products that, upon termination of their activity, must be able to convert into products that are harmless to the environment.

Fulfillment of this requirement is possible by changing the technological parameters in the process management and the change of so-called. Auxiliary substances added at certain stages in the production process. The aim is to prevent the formation of harmful substances and to return to production as much waste as possible, which is achieved by recycling.

Real-time Analysis for Pollution Prevention

Analytical methodologies need to be further developed to allow for real-time, in-process monitoring and control prior to the formation of hazardous substances. Traditional analytical chemistry implies large amounts of sample for analysis, abundant use of solvents and energy. With the development of new methods and precision mobile instruments, it is possible that the analyzes work with a small sample size at the sampling site and with much less solvent.

The principle of Real-Time Analysis for Pollution Prevention requires further development of analytical methodology to enable real-time monitoring of the chemical production process with the aim of preventing the formation of dangerous substances, i.e. it is necessary to constantly monitor the production process at each stage would prevent the occurrence of errors that could lead to the emergence of dangerous substances, harmful to the environment and human health.

Inherently Safer Chemistry for Accident Prevention

Substances and the form of a substance used in a chemical process should be chosen to minimize the potential for chemical accidents, including releases, explosions, and fires. The Twelfth Principle of Green Chemistry is the principle of Inherently Safer Chemistry for Accident Prevention. The basic requirement is to reduce the use of substances in chemical processes that can cause adverse effects (explosion, fire and harmful vapor). An example is today the increasing use of supercritical CO_2 that replaces organic solvents and which, unlike organic solvents, is not toxic or explosive and is environmentally acceptable.

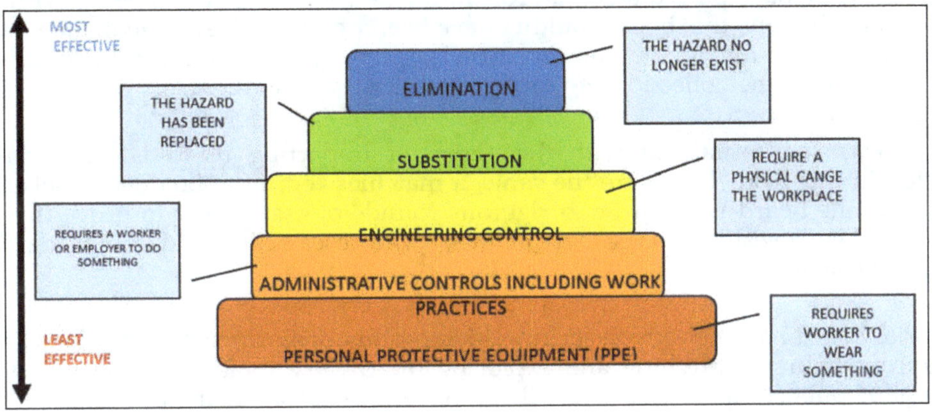

Hierarchy of security control.

Safety can be defined as a control of known hazards by achieving an acceptable level of risk and is achieved at several levels of the lowest use of Personal Protective Equipment. Then it follows the level of Administrative and Work Practice Controls) and implies establishing effective procedures, rotating work tasks, adjusting work schedules so that workers are not over-exposed to the impact of dangerous chemicals, etc. The next higher level of security control is the expert Engineering Controls, which implies the implementation of physical process change to reduce contact with hazardous chemicals, isolate the process, use wet methods to reduce dust formation, ventilation, digestion, etc. The highest level of safety control is achieved by eliminating or replacing the procedure with safer alternatives.

References

- What-is-green-chemistry, greenchemistry: acs.org, Retrieved 20 August, 2020
- Sheldon, R. A. (2007). "The E Factor: Fifteen years on". Green Chemistry. 9 (12): 1273. doi:10.1039/B713736M
- Review-of-12-Principles-of-Green-Chemistry-in-Practice-319273400: researchgate.net, Retrieved 2 July, 2020
- Sheldon R. A. (2000). "Atom efficiency and catalysis in organic synthesis" (PDF). Pure and Applied Chemistry. 72 (7): 1233–1246. doi:10.1351/pac200072071233

Permissions

All chapters in this book are published with permission under the Creative Commons Attribution Share Alike License or equivalent. Every chapter published in this book has been scrutinized by our experts. Their significance has been extensively debated. The topics covered herein carry significant information for a comprehensive understanding. They may even be implemented as practical applications or may be referred to as a beginning point for further studies.

We would like to thank the editorial team for lending their expertise to make the book truly unique. They have played a crucial role in the development of this book. Without their invaluable contributions this book wouldn't have been possible. They have made vital efforts to compile up to date information on the varied aspects of this subject to make this book a valuable addition to the collection of many professionals and students.

This book was conceptualized with the vision of imparting up-to-date and integrated information in this field. To ensure the same, a matchless editorial board was set up. Every individual on the board went through rigorous rounds of assessment to prove their worth. After which they invested a large part of their time researching and compiling the most relevant data for our readers.

The editorial board has been involved in producing this book since its inception. They have spent rigorous hours researching and exploring the diverse topics which have resulted in the successful publishing of this book. They have passed on their knowledge of decades through this book. To expedite this challenging task, the publisher supported the team at every step. A small team of assistant editors was also appointed to further simplify the editing procedure and attain best results for the readers.

Apart from the editorial board, the designing team has also invested a significant amount of their time in understanding the subject and creating the most relevant covers. They scrutinized every image to scout for the most suitable representation of the subject and create an appropriate cover for the book.

The publishing team has been an ardent support to the editorial, designing and production team. Their endless efforts to recruit the best for this project, has resulted in the accomplishment of this book. They are a veteran in the field of academics and their pool of knowledge is as vast as their experience in printing. Their expertise and guidance has proved useful at every step. Their uncompromising quality standards have made this book an exceptional effort. Their encouragement from time to time has been an inspiration for everyone.

The publisher and the editorial board hope that this book will prove to be a valuable piece of knowledge for students, practitioners and scholars across the globe.

Index

www.ingramcontent.com/pod-product-compliance
Lightning Source LLC
LaVergne TN
LVHW080626210926
848485LV00003B/218

* 9 7 8 1 6 3 9 8 9 1 8 0 1 *